Science, Religion and Belief

American University Studies

Series VII
Theology and Religion

Vol. 106

PETER LANG
New York • San Francisco • Bern • Baltimore
Frankfurt am Main • Berlin • Wien • Paris

Rogers B. Miles

Science, Religion and Belief

The Clerical Virtuosi of the Royal Society of London, 1663-1687

PETER LANG
New York • San Francisco • Bern • Baltimore
Frankfurt am Main • Berlin • Wien • Paris

Library of Congress Cataloging-in-Publication Data

Miles, Rogers Blood.
Science, religion, and belief : the clerical virtuosi of the Royal Society of London, 1663-1687 / Rogers B. Miles.
p. cm. — (American university studies. Series VII, Theology and religion ; v. 106)
Includes bibliographical references .
1. Church of England—Clergy. 2. Anglican Communion—England—Clergy. 3. Royal Society (Great Britain)
4. Apologetics—17th century. I. Title. II. Series.
BX5175.M54 1992 283'.092'242—dc20 91-16763
ISBN 0-8204-1564-2 CIP
ISSN 0740-0446

Die Deutsche Bibliothek-CIP-Einheitsaufnahme

Miles, Rogers B.:
Science, religion, and belief : the clerical virtuosi of the Royal Society of London, 1663-1687 / Rogers B. Miles.—New York; Berlin; Bern; Frankfurt/M.; Paris; Wien; Lang, 1992
(American university studies : Ser. 7, Theology and religion ; Vol. 106)
ISBN 0-8204-1564-2
NE: American university studies/07

The paper in this book meets the guidelines for permanence and durability of the Committee on Production Guidelines for Book Longevity of the Council on Library Resources.

Printed in the United States of America.

CONTENTS

ACKNOWLEDGMENTS

When the historian, Perry Miller, republished his doctoral dissertation in 1953, he praised his dissertation adviser for daring to give his students sufficient rope to hang themselves if this be their resolve. I wish to thank both my advisers, Horton M. Davies and John F. Wilson, for the leash I was given during the course of my dissertation to pursue the research that forms the body of this study. Their forbearance was an act of faith which I hope is now rewarded.

Two libraries became my home during the course of my research. To the staffs of the Folger Shakespeare Library and the Library of Congress I owe a special debt of thanks for the assistance they bestowed. I am grateful to those friends who helped to clarify my thinking. Steve Ackerman, Margreta DeGrazia, John Kenney, Elizabeth Kessel, Pamela Long, Nancy Matthews, Richard Popkin, Jeffrey Stout, Thorell Tsomondo, Jean Veta, and Dewey Wallace made invaluable contributions to the shaping of my ideas. If I mentioned all the friends who encouraged me along the way, the list would be too long. They know who they are and they have my appreciation.

Special mention must be made of the late Commander Marjorie E. Wright whose generous bequest helped bring this book to print. I also thank Jenny Russell for the care and the forethought with which she transformed my manuscript into camera ready copy.

The debt I owe my family is greater than I can express. Throughout the entire effort, the backing of my parents—both moral and financial—sustained my will to persevere. Though my mother lived to see the completion of the dissertation, she is not here to witness its publication. This book is dedicated to her memory.

Rogers B. Miles
Walla Walla, Washington

Elizabeth Blood Miles

1914 – 1987

– 1 –

INTRODUCTION

The following is a study of fifty-three clergymen who were elected to the Royal Society between 1663 and 1687. Such a study is of interest for what it reveals about the nature of the early Royal Society and the reasons which led its members to seek election to its rolls. The Royal Society of London both contributed to and was a social manifestation of a growing interest in natural philosophy among the educated classes of seventeenth-century Europe. The goal of the Society—in the words of Henry Oldenburg, the Society's first secretary—was "to cultivate knowledge of nature and useful arts by means of observation and experiment, and to promote them for the safeguarding and convenience of human life."[1] Every member was expected to contribute to this goal, for the Society aimed to be not the honorary body that it eventually became nor the London social club that it threatened to become, but a research institution where fellows would work in common in the pursuit of natural knowledge. That this did not come to pass was an unintended consequence of its means of raising money to finance its operations.[2]

In contrast to its rival, the Academie Royale des Sciences, the Royal Society raised its funds from gifts and levies upon its members. Although Charles II declared himself the Society's founder and patron in approving its second charter in 1663, his support was not sufficient to free the Society from its dependence upon the income of its members. When its members were not delinquent in the payment of their dues, the revenue produced barely covered its expenses. To generate more income the Society expanded its rolls well beyond the fifty-five originally envisaged by its founders.[3] This expansion added many who had only a superficial interest in the work of the Society.

This may help to explain in part why over a quarter of the clerics who are the subject of this study took no part in the Society after the date of their election. Much of this deadwood entered the Society during the 1660's when election to the Society still carried some prestige. Seven bishops, for example, joined the

Society in this period.[4] Only one was anything more than an ex officio member. The other six did not attend any meeting beyond their first, nor did they publish anything of scientific import. Such members did not improve the Society's balance sheets since their support remained contingent upon the Society's ability to pander to their interests. Once they tired of its proceedings, they stopped coming to its meetings and defaulted on their subscriptions. Only sixteen of the fifty-three clerics were active to any extent in the proceedings of the Society throughout the period they were fellows.[5] Another twelve displayed some interest after the date of their election, but few continued to be active any longer than three years.[6]

For those who did not fall away, the Society's weekly meetings provided a forum in which to continue the discussion of learned topics they had enjoyed at university.[7] Nearly all the clerics who were active in the Society in the period of this study were university educated. All but one held advanced degrees that required the study of science as a part of their curriculum.[8] Many had lived in colleges where coteries existed that were interested in science. Seven clerics had been residents of Trinity College, Cambridge during the 1650's and 1660's, a period when the college had been the center of Cambridge science. Five had been members of the Oxford group which met at Wadham College during the 1650's. At least one had been in residence at Trinity College, Oxford in the 1640's when that college had experienced a renewal of interest in the sciences. Another cleric had resided in Merton College, Oxford under the wardenship of William Harvey.[9]

The friendships that they formed during their years at university were responsible in part for their recruitment to the Society. John Wilkins nominated seven clerics to the Society. At least two he had known during his years in Oxford as warden of Wadham College. They had been members of the club over which he had presided. Another two he may have known from his year in Cambridge when he was president of Trinity College.[10]

This study does not consider whether religion disposed a cleric to engage in science or seek election to the Society. It is the opinion of the author that religion in the seventeenth century was less a motive for doing science than a way of justifying scientific activity. If religion had been the motive that Robert K. Merton postulated, we would not expect to see the diversity of outlook that we see among the fellows of the Royal Society.[11] Even Charles Webster who has championed the importance of Puritanism to the growth of English science has conceded the Royal Society was not the predominantly Puritan body that some historians have imagined.[12]

This is not to say that there is no correlation between the science of the Royal Society and the religion of its members. Charles Webster has observed that Puritans and Anglicans developed contrasting attitudes toward nature that tended to coincide with their political and religious beliefs. The study of agriculture, for example, was a Puritan preoccupation whereas experimental physiology attracted Anglicans and royalists. In making such correlations it is important not to ignore the welter of opinions subsumed by party labels like Puritan or Anglican. Among those labeled Puritan were left-wing Puritans like Samuel Hartlib and John Dury who saw the significance of science in terms of an eschatology which promised the eventual amelioration of the human condition by the recovery of Adamic knowledge. The same label also included right-wing Puritans like John Wilkins and John Wallis who saw the significance of science in terms of the evidence it gave of the author of creation.[13] There was considerable interchange between proponents of different viewpoints and even occasional collaboration in the pursuit of common goals. Michael Hunter cites the example of the royalist John Evelyn who was not averse to corresponding with a Puritan like Samuel Hartlib.[14]

A similar diversity also characterized the assumptions about knowledge held by members of the Society. Historians have often written as if a specific theory of knowledge inspired the Society's methodology. Thus Margery Purver has sought to argue that the Society looked to Bacon for the source of its methodology.[15] Others like Laurens Laudans believe it took its methodology from the philosophy of Descartes.[16] Still others like Richard Popkin and Henry Van Leeuwen claim that mitigated skepticism was its source of inspiration.[17]

According to Popkin and Van Leeuwen, mitigated skepticism accepted the Pyrrhonist critique of knowledge. It denied there were any criteria by which a person could be certain that their knowledge corresponded with reality itself. In the absence of such criteria it sought to justify human knowledge upon the grounds of probability. They suggest that mitigated skepticism is behind the Society's suspicion of opinions whose only backing is authority and its oft-stated warning against regarding its own conclusions as anything more than tentative. We shall see in Chapter 5 that a cleric could endorse the Society's methodological caution without subscribing to the skepticism which supposedly inspired it. Such a cleric could acknowledge the fallibility of human knowledge, and still affirm that certain knowledge was theoretically possible.

In Chapter 4 we will examine how doubts about the certainty to which human knowledge could attain may have prompted those who held them to redefine belief and the reasons that justified it. By focusing on the ways clerics defined the term "belief," a correlation can be established between their stand on the problem of certainty and their success in answering critics of the mysteries of Christianity.

The next two chapters set the stage for the discussions in four and five. Chapter 2 is a prosopography of the fifty-three clerics who were elected to the Society between 1663 and 1687. It charts the course of their careers, assesses their contribution to science and theology, and investigates the factors that led to their election to the Royal Society. Chapter 3 examines the extent of their activity in the life of the Society.

NOTES

1 Henry Oldenburg to Dr. Eccard Leichner, 23 September 1663, *The Correspondence of Henry Oldenburg*, trans. and ed. A. Rupert Hall and Marie Boas Hall, 10 vols. (Madison: The University of Wisconsin Press, 1965-75), 2:111.

2 See R. K. Bluhm, "Remarks on the Royal Society's Finances," *Notes and Records of the Royal Society of London* 13 (November 1958):82-103; Michael Cyril William Hunter, "The Social Basis and Changing Fortunes of an Early Scientific Institution: An Analysis of the Membership of the Royal Society, 1660-1685," *Notes and Records of the Royal Society of London,* 31 (July 1976):16-19; and Michael Cyril William Hunter, *Science and Society in Restoration England* (Cambridge: Cambridge University Press, 1981), pp. 40-41.

3 At the third meeting of the yet unnamed and unchartered Royal Society, the decision was made to set the membership limit at fifty-five. For the minutes of the meeting, 12 December 1660, see Thomas Birch, *The History of the Royal Society of London for Improving of Natural Knowledge From its First Rise. In which The most considerable of those Papers communicated to the Society, which have hitherto not been published, are inserted in their proper order, as a supplement to The Philosophical Transactions*, with an introduction by A. Rupert Hall and additional bibliographical note by Marie Boas Hall, 4 vols. (London: n.p., 1756-57; reprint ed., New York: Johnson Reprint Co., The Sources of Science, no. 44, 1968), 1:5. By 1663, when the Council released their tally of fellows elected there were one hundred fifteen names. See Birch, *History,* 1:239-40. In 1667, Thomas Sprat's list of fellows contained 191 names. See Thomas Sprat, *History of the Royal-Society of London, For the Improving of Natural Knowledge,* ed. Jackson I. Cope and Harold Whitmore Jones (London: Printed by T.R. for J. Martyn at the Bell without Temple-bar, and J. Allestry at the Rose and Crown in Duck-Lane, Printers to the Royal Society, 1667; reprint ed., St. Louis: Washington University Studies, 1958), pp. 431-33. Between 1663 and 1687, four hundred twenty-nine names were registered on the rolls. This figure is compiled from the Halls' index of fellows given in the introduction to Birch, *History*, 1:xxxiii-xlv.

4 The bishops were the following: John Dolben, bishop of Rochester, future archbishop of York, Humphrey Henchman, bishop of London; Benjamin Laney, bishop of Lincoln; George Morley, bishop of Winchester; Gilbert Sheldon, archbishop of Canterbury; Richard Stearne, archbishop of York; and Edward Wetenhall, bishop of Cork and Ross.

5 Using Michael Hunter's categories, ten of the sixteen can be classified as "active," "fairly active," or "very active" in the Society. They were St. George Ashe, John Beale, John Flamsteed, Thomas Gale, William Holder, William Perrey, Nathaniel Vincent, John Wallis, John Wilkins, and Robert Wood. Another six were "barely" or "slightly" active in the proceedings of the Society. They were Ralph Bathurst, Joseph Glanvill, John Ray, Thomas Smith, Edward Wetenhall, and Sir George Wheler. See Hunter's "Catalogue of fellows 1660-1685" in Hunter, "Membership," pp. 97-114. For Hunter's discussion of his categories, see Hunter, "Membership," pp. 47-49.

6 Of the twelve who showed some interest after the date of their election, three were "active," "fairly active," or "very active" in the Society at the height of their interest. They were John Mapletoft, John Pell, and Seth Ward. The other nine were never more than "barely" or

"slightly" active in the proceedings of the Society. They were James Arderne, William Harrington, Anthony Horneck, Daniel Milles, William Payne, John Pearson, Thomas Pigot, Isaac Vossius, and Edward Waterhouse. Waterhouse, for example, attended six meetings in the space of seven months before lapsing in his attendance. Others were little better. Mapletoft's last attendance at a meeting of the Society was only three years after his first. The only exceptions among the twelve were Pell and Ward. Pell was active through the 1660's and Ward remained active until the middle 1670's. These statistics are compiled from Hunter's "Catalogue of fellows 1660-1685" in Hunter, "Membership," pp. 97-114, and Birch's *History*, using Gail Scala's index. See Gail Ewald Scala, "An Index of Proper Names in Thomas Birch, *The History of the Royal Society* (London, 1756-1757)," *Notes and Records of the Royal Society of London* 28 (April 1974):263-329.

7 See Robert Gregg Frank, Jr., "Science, Medicine and the Universities of Early Modern England: Background and Sources," *History of Science* 11 (1973):199.

8 Only Edward Waterhouse among the fellows who were active was not a product of the universities. Twenty-six of the twenty-eight clerics who were active in the Society held M.A.'s from Oxford or Cambridge at the time of their election. Isaac Vossius earned the same degree from a foreign university. Their M.A.'s are significant, for they attest to their exposure to geometry, astronomy, and natural philosophy. Robert G. Frank has observed that these three subjects formed almost forty percent of the M.A. curriculum. See Frank, "Science," 202.

9 For the importance of coteries in drawing students to the sciences, see Frank, "Science," 257-58. Thomas Gale, John Mapletoft, John Pearson, William Perrey, John Ray, Edward Wetenhall, and John Wilkins resided in Trinity College, Cambridge during 1650's and 1660's. Joseph Glanvill, John Wallis, Seth Ward, John Wilkins, and Robert Wood were members of one of the precursors of the Royal Society, The Oxford Experimental Philosophy Club, which met at Wadham College in the mid-1650's. The one cleric who was resident at Trinity College, Oxford in the 1640's was Ralph Bathurst. Robert Wood was the cleric who resided at Merton College when William Harvey was its warden.

10 John Wilkins nominated William Harrington, Henry More, Samuel Parker, John Ray, Thomas Sprat, Edward Waterhouse, and Samuel Woodford to the Royal Society. Parker and Sprat had previously been members of the Oxford Experimental Philosophy Club. Ray had resided at Trinity College, Cambridge when Wilkins was its president. For the importance of friendship as a factor in recruitment, see my discussion in chapter 1 and Frank, "Science," 263. For a broader discussion of how the university environment helped to foster ties of friendship, see Frank, "Science," 257-61 and Robert Gregg Frank Jr., *Harvey and the Oxford Physiologists: A Study of Scientific Ideas and Social Interaction* (Berkeley: University of California Press, 1980), pp. 57-62.

11 See Robert K. Merton, *Science, Technology, and Society in Seventeenth Century England*, (New York: Howard Fertig, 1970), particularly pp. 80-83, for his discussion of "motive forces" and pp. 112-115, for his analysis of "Puritan elements" within the Royal Society. Merton's study first appeared in *Osiris* 4 (1938):360-632.

12 It is Webster's opinion in "The Origins of the Royal Society," *History of Science* 6 (1967):125-26, that "no single pre-Restoration group was responsible for the Royal Society." See also Charles Webster, *The Great Instauration: Science, Medicine, and Reform, 1626-1650* (London: Duckworth, 1975), pp. 93-95, for Webster's analysis of the religious motivations of the active nucleus of the Society, and Ibid., pp. 495-96, for Webster's discussion of the heterogeneity of Restoration science. Webster does not conclude as Lotte Mulligan has done that the Society must have been Anglican if it was not Puritan. He criticizes her attempt

to assign its members to either party on the basis of their affiliations in the Interregnum period. See Lotte Mulligan, "Civil War Politics, Religion and the Royal Society," in *The Intellectual Revolution of the Seventeenth Century*, ed. Charles Webster (Boston: Routledge & Kegan Paul, Ltd., 1974), pp. 317-339. For Webster's criticisms of Mulligan, see Webster's introduction to *Intellectual Revolution*, pp. 20-21. Hunter's comments are also germane. See Hunter, "Membership," pp. 10, 15; and Hunter, *Science*, 115-117.

Among the historians who have stressed the Puritan background of the Royal Society are Dorothy Stimson in *Scientists and Amateurs* (London: Sigma Books Ltd., 1949), pp. 34-35, 39, 71, and "Puritanism and the New Philosophy in Seventeenth-Century England," *Bulletin of the Institute of the History of Medicine* 3 (May 1935):321-34; and Christopher Hill in the following: "Puritanism, Capitalism and the Scientific Revolution," in *The Intellectual Revolution of the Seventeenth Century*, ed. Charles Webster (Boston: Routledge & Kegan Paul, Ltd., 1974), pp. 243-53; "The Intellectual Origins of the Royal Society, London or Oxford?" *Notes and Records of the Royal Society of London* 23 (1968):144-56; and *Intellectual Origins of the English Revolution* (London: Panther Books, 1972), pp. 105-8, 125-30.

Barbara Shapiro has come to wonder whether the existence of such connections between Anglicans and Puritans does not undermine the notion that Puritanism or Anglicanism can adequately comprehend the relation that existed between science and religion in seventeenth-century England. She writes that "it is particularly the absence of research on non-Laudian, moderate Anglicans, combined with the extensive work on the puritan movement, that has obscured the fact that moderation *qua* moderation, not the puritanism of puritan moderates, provides the key link between science and religion." See Barbara J. Shapiro, "Latitudinarianism and Science in Seventeenth-Century England," in *The Intellectual Revolution of the Seventeenth Century*, ed. Charles Webster (Boston: Routledge & Kegan Paul, Ltd., 1974), p. 290. Her thesis gains plausibility from current skepticism concerning the usefulness of terms like Puritanism and Anglicanism for grasping the politics of the period. See Michael G. Finlayson, *Historians, Puritanism, and the English Revolution: The Religious Factor in English Politics Before and After the Interregnum* (Buffalo: University of Toronto Press, 1983), particularly pp. 3-8, 42-76, 159-162.

13 Webster, *Instauration*, pp. 498-99.

14 Hunter, *Science*, p. 26. See also Webster, *Instauration*, pp. 97-98.

15 See Margery Purver, *The Royal Society: Concept and Creation* (London: Routledge and Kegan Paul, 1967), chaps. 2 and 3.

16 See Laurens Laudan, "The Clock Metaphor and Probabilism: The Impact of Descartes on English Methodological Thought, 1650-65," *Annals of Science* 22 (June 1966):73-104.

17 See Richard H. Popkin, *The History of Scepticism from Erasmus to Spinoza* (Berkeley: University of California Press, 1979), pp. 146-47. See also Henry G. Van Leeuwen, *The Problem of Certainty in English Thought: 1630-90* (The Hague: Martinus Nijhoff, 1963), and Robert Todd Carroll, *The Common-sense Philosophy of Religion of Bishop Edward Stillingfleet 1635-1699* (The Hague: Martinus Nijhoff, 1975).

– 2 –

THE CLERICAL VIRTUOSI

This chapter focuses on the lives and reputations of the fifty-three clergymen who were fellows of the Royal Society of London between 1663 and 1687. Only eight of the fifty-three receive no mention in the *Dictionary of National Biography.* This fact is significant because it indicates that the clerical fellows of the Royal Society were men of note among their contemporaries. Some were known chiefly for their work in natural philosophy. Others were known for their learning. Most, however, gained prominence by their preferment. Of the ten who receive articles in the *Dictionary of Scientific Biography*, only three were not prominent in church or university circles. This chapter examines the careers of these fifty-three men of the cloth and the factors that led to their election to the Royal Society.

Of the four hundred fourteen fellows elected to the Royal Society of London between May 20, 1663, and December 31, 1687, fifty-one were ordained at the time of their election while another two were ordained during the course of their membership. Table 2.1 lists their names and the dates of their election. No claim is made that this list includes every cleric who was a fellow in this period. Some of the "foreign fellows" elected to the Society were clerics. None are included here except Isaac Vossius, who emigrated to England six years after his election. The others were excluded because of their geographical isolation. Their separate membership status reflected this isolation. As "foreign fellows," they were not required to pay dues in support of the Society since their distance from London prevented them from fully participating in its activities. Though many carried on a lively correspondence with the Society and some even had occasion to attend one or more of its meetings on visits to England, they had little impact upon the course of its proceedings. Excluded too are those clerics who were elected under the first charter, but who were not reelected under the second.

TABLE 2.1

THE CLERICAL VIRTUOSI OF THE ROYAL SOCIETY, 1663-1687

Name	Birth Date	Death
James Arderne	1636	1691
St. George Ashe	1658?	1718
Thomas Baker	1625?	1690
Isaac Barrow	1630	1677
Ralph Bathurst	1620	1704
John Beale	c. 1613	c. 1682
Edward Bernard	1638	1696
Gilbert Burnet	1643	1715
Edmund Castell	1606	1686
Henry Clerke	1622?	1687
Edward Cotton	1616	1675
John Dolben	1625	1686
John Flamsteed	1646	1719
Thomas Gale	1635?	1702
Joseph Glanvill	1636	1680
William Harrington	1623?	1697?
Humphrey Henchman	1592	1675
William Holder	1616	1698
Anthony Horneck	1641	1697
Charles Hotham	1615	1672
Henry Jenkes	?	1697
Benjamin Laney	1591	1675
John Mapletoft	1631	1721
Daniel Milles	?	1689
Henry More	1614	1687
George Morley	1597	1684
Samuel Parker	1640	1688
William Payne	1650?	1696
John Pearson	1613	1686
John Pell	1611	1685
William Perrey	1650	1713
Thomas Pigot	1657	1686

TABLE 2.1 — *Continued*

Name	Birth Date	Death
Francis Potter	1594	1678
John Ray	1627	1705
John Rogers	1647	1715
Gilbert Sheldon	1598	1677
Thomas Smith	1638	1710
Thomas Sprat	1635	1713
Richard Stearne	1596?	1683
John Tillotson	1630	1694
Nathaniel Vincent	?	1722
Isaac Vossius	1618	1689
John Wallis	1616	1703
Seth Ward	1617	1689
Edward Waterhouse	1619	1670
Edward Wetenhall	1636	1713
Sir George Wheler	1650	1723
John Wilkins	1614	1672
Roger Williams	?	1702/03
Robert Wood	1622?	1685
Samuel Woodford	1636	1701
Benjamin Woodroffe	1638	1711
William Wotton	1666	1727

SOURCES: *Dictionary of National Biography From the Earliest Times to 1900*, 22 vols. (Oxford: Oxford University Press, 1885-1900); Catalogue of Fellows 1660-1685 in Michael Cyril William Hunter, "The Social Basis and Changing Fortunes of an Early Scientific Institution: An Analysis of the Membership of the Royal Society, 1660-1685," *Notes and Records of the Royal Society of London* 31 (July 1976):97-114; Joseph Foster, *Alumni Oxonienses,* 4 vols. (Oxford: Parker & Co., 1891-92); John Le Neve, *Fasti Ecclesiae Anglicanae, 1541-1857*, 4 vols. (London: The Athlone Press for The Institute of Historical Research, 1969-75); A. G. Matthews, *Calamy Revised* (Oxford: Clarendon Press, 1934) and Idem, *Walker Revised* (Oxford: Clarendon Press, 1948); Officers and Fellows of the Royal Society 1663-1687 in Thomas Birch, *The History of the Royal Society,* 4 vols. (London: n.p., 1756-57; reprint ed., with an introduction by A. Rupert Hall and additional bibliographic note by Marie Boas Hall, The Sources of Science, no. 44, New York: Johnson Reprint Co., 1968), 1:xxxiii-xlv; John Venn, *Biographical History of Gonville and Caius College, 1349-1897*, 4 vols. (Cambridge: Cambridge University Press, 1897-1912), vol. 1: *Admissions 1349-1713*; John Venn, and J. A. Venn, *Alumni Cantabrigiensis*, 4 vols. (Cambridge: Cambridge University Press, 1922-54).

It is a sign of their prominence in Restoration society that only two of the names listed in Table 2.1 escape positive identification. What proved considerably more difficult than establishing their identities was confirming their clerical status. Unless the *Dictionary of National Biography* or some other biographical source confirmed the fact of their ordination, it could not be verified without a search of episcopal registers. This was not undertaken. In lieu of such a search, an attempt was made to infer ordination using indirect evidence. Ordination could safely be assumed in cases where a fellow was a bishop, parson, or curate, but if a fellow did not hold ecclesiastical office, the fact of his ordination had to be inferred from evidence that was considerably less reliable.

The probability that a fellow took orders was high, for example, if he earned an M.A., and a certainty if he had received a B.D. or a D.D. It should be noted, however, that there were no degree requirements for ordination in the seventeenth century and that an M.D. or LL.D. did not preclude the possibility of ordination. The career of Ralph Bathurst is a case in point. Ordained in 1644, he pursued medicine in the 1650's because of his distaste for the Puritanism of the Oxford theological faculty.[1]

Another sign of ordination was the holding of college fellowships. Most fellowships in unreformed Oxford and Cambridge required ordination for their tenure, though this rule was often bent and sometimes broken. Isaac Barrow, for example, did not take orders until 1659, even though he had been a fellow at Trinity College, Cambridge since 1649. Many colleges also offered fellowships in medicine and civil law which did not require ordination. The recipients of these, unfortunately, are not distinguished in the alumni lists compiled by Foster and Venn.[2] For reasons such as these, there may be clerics who escaped inclusion in the list given in Table 2.1.

Birth and Learning

It was the opinion of Henry Oldenburg, the Society's first secretary, that the Royal Society contained "many men distinguished by their birth, learning, and wit."[3] This certainly appears to be the case if the reputations of those who are listed in Table 2.1 are a fair sample of the membership as a whole. Oldenburg's statement supports the view that the social status of members of the Royal Society was similar to that of a gentlemen's club. Like many clubs the Royal Society had a broad social purpose, the improvement of natural knowledge, and yet like many clubs, it devoted as much energy to asserting its claims to exclusiveness and importance as it did to pursuing its stated goals. Though anyone could be proposed for membership, few were admitted who could not boast birth, learning, or wit. The election of fellows with these qualities not only helped to cultivate an image which made membership desirable, it also provided those casting the votes an occasion for reaffirming their own good opinion of themselves.

Of the three desirable qualities, birth commanded the greatest respect. The charter of the Royal Society prescribed that persons of noble birth be admitted upon being proposed. They were termed "honorary fellows." They did not have to submit to a vote nor were they required to pay subscriptions in support of the Society's activities. Five clerical virtuosi were admitted as honorary fellows. They were Gilbert Sheldon, Richard Stearne, Humphrey Henchman, George Morley, and John Dolben. None were of noble blood, but as lords of the Church, they were members of the nobility by virtue of their office. When Sheldon, Stearne, and Henchman joined the Society in March 1665, Sheldon was archbishop of Canterbury, Stearne was archbishop of York, and Henchman was bishop of London. The bishop of Winchester, George Morley, joined a year later in November of 1666.

John Dolben, however, was not yet a lord of the Church when he was invited to join the Society as a honorary fellow. Perhaps his admission to the Society at the same time as his uncle, the Archbishop of Canterbury, accounts for this anomaly.[4] Dolben became a bishop within a year of joining the Society, and succeeded Stearne at York upon the latter's death in 1683. Even though Benjamin Laney was bishop of Lincoln at the time of his election, the Society did not acknowledge his position with an honorary fellowship. The oversight was at least partly Laney's fault. Had he attended a meeting after being proposed, it is likely that he would have been inducted into the Society as an honorary fellow. Nevertheless, the Society often failed to be consistent in its extension of this privilege. Edward Wetenhall was bishop of Cork and Ross and yet his status in the Society was that of a regular fellow.

Eight fellows were elected who were destined to become bishops after joining the Society. They were John Tillotson, John Wilkins, John Pearson, Seth Ward, Gilbert Burnet, Samuel Parker, Thomas Sprat, and St. George Ashe. John Tillotson was a lecturer at St. Lawrence Jewry when he was elected to the Society in 1672. Twenty years later he was archbishop of Canterbury. His father-in-law, John Wilkins, was dean of Ripon when he helped form the group in 1660, which subsequently became the Royal Society. Despite the fact that he was married to Cromwell's sister, John Wilkins was made bishop of Chester in 1668. When Wilkins died in 1672, another clerical virtuoso succeeded him at Chester, the noted divine and theologian, John Pearson. Seth Ward, mathematician and future bishop, was vicar of St. Lawrence Jewry in 1661, when he joined the group that would become the Royal Society. When Ward became bishop of Exeter in 1662, he saw to it that his friend, John Wilkins, succeeded him as vicar of St. Lawrence Jewry. Ward was translated from Exeter to Salisbury in 1667. The historian, Gilbert Burnet was a young man visiting London on his way to Rotterdam when he was proposed for membership by his friend and fellow Scotsman, Sir Robert Moray. Burnet eventually became bishop of Salisbury after the Glorious Revolution. Samuel Parker was residing at Trinity College, Oxford when he was elected to the Society in 1666. He left Oxford in 1667, to become chaplain to Archbishop Sheldon, only to return as bishop in 1686. Though a fellow of Wadham College, Oxford, Thomas Sprat, future author of the *History of the Royal Society*, had already gone down to London, when he was elected to the Society in April of 1663. Sprat rose to become bishop of Rochester in 1684. St. George Ashe was secretary of the Dublin Philosophical Society when he was invited to join the Royal Society in 1686. In 1695, Ashe was elevated to the bishopric of Cloyne. He was translated to Clogher in 1697, and to Derry in 1717.

Learning, even more than birth, was the quality most highly esteemed in the circles from which the Society did its recruiting. The number of academicians among the clerical virtuosi was, in part, a reflection of this bias. Four of the bishops who joined the Royal Society had been academicians during the reign of Charles I. Sheldon had been warden of All Souls, Oxford; Stearne had been master of Jesus College, Cambridge; and Laney had been master of Pembroke Hall, Cambridge. Morley, a canon of Christ Church, Oxford, until he was suspended in 1647, was awarded the deanship of his college at the Restoration.

Four of the eight clerical fellows destined to become bishops first made their mark in academe before being elevated to the episcopal bench. John Pearson, the foremost English patristic scholar after Archbishop Ussher, succeeded Stearne as master of Jesus College, Cambridge in 1662. He went on to become master of Trinity College, Cambridge and Lady Margaret professor of Divinity, before succeeding Wilkins as bishop of Chester. St. George Ashe was

Donegal lecturer and professor of mathematics at Trinity College, Dublin, when he was forced to flee Ireland during the Williamite Revolution. In 1692, he returned to Trinity as its provost, where three years later he received his appointment to the bishopric of Cloyne. John Wilkins had been warden of Wadham College, Oxford during most of the Interregnum when he accepted the mastership of Trinity College, Cambridge, in 1659. His tenure there was brief. He was forced to resign his headship at the Restoration. Seth Ward had been Savilian professor of astronomy when he became president of Trinity College, Oxford in 1659. He too resigned his headship at the Restoration.

The ranks of the bishops and bishops-to-be do not exhaust the number of clerical virtuosi who were prominent academicians. Ralph Bathurst, John Wallis, Isaac Barrow, Henry Clerke, Edward Bernard, and Thomas Gale had academic careers which were equally illustrious though they were never raised to the episcopal bench. Of the group, only Gale did not spend his entire career in either Oxford or Cambridge. Gale had been in Cambridge for seventeen years when, in 1672, he resigned his Regius chair of Greek and his fellowship at Trinity College, and accepted the high mastership of St. Paul's School in London. As high master of St. Paul's, Gale continued to pursue his historical and classical studies until he was nominated dean of York, five years before his death in 1702.

Gale left Trinity one year before the mathematician, Isaac Barrow, assumed its mastership. The mastership of Trinity College was the culmination of Barrow's career at Cambridge. He commenced his climb up the rungs of university preferment when he was awarded the Regius professorship of Greek at the Restoration. In 1662, he was appointed Gresham professor of Geometry. The latter position required weekly trips to London, and it was during this period that he joined the Royal Society. He resigned both positions in 1663, to become the first Lucasian professor of mathematics, which he in turn resigned for the mastership of Trinity in 1673.

Barrow's rapid rise in preferment was matched by Ralph Bathurst's longevity in office. Bathurst migrated to Trinity College, Oxford, in 1637, when he was seventeen. He obtained a fellowship there when he was twenty and was awarded his bachelor and doctor of medicine degrees in his early thirties. He became president of Trinity when he was forty-four and forty years later he was still its president when he died in office. When Henry Clerke joined the Royal Society in October of 1667, he had been a fellow at Magdalen College, Oxford, for twenty-five years, and for twelve of those years he had been Tomlins lecturer in anatomy. He was destined to rise even further. In 1672, he became president of Magdalen College, and was its president until his death in 1687.

John Wallis never rose to a college headship though his tenure in the Savilian chair of geometry topped Bathurst's term as president. The mathematician held the Savilian chair from 1649 to 1703, a span of fifty-four years. His student, the noted oriental scholar, Edward Bernard, was another who spent nearly his entire career at Oxford. Bernard was named the Savilian professor of astronomy at Oxford the same year that he was elected to the Royal Society, and he held the chair for the next eighteen years before resigning it in 1691, for a living in Berkshire. The semitic scholar, Edmund Castell, was professor of Arabic in Cambridge for twenty-two years, from 1664 until his death in 1686.

Bathurst, Wallis, Barrow, Clerke, Bernard, and Castell were fortunate to have received preferment. Many of the clerical virtuosi who chose to spend their lives in Oxford or Cambridge advanced no further than a college fellowship. There were only a limited number of places in either university, and seniority and patronage increasingly determined who would get them. Benjamin Woodroffe, for example, was canon of Christ Church College, Oxford, for thirty-nine years from 1672 to 1711. Thomas Smith tried to be appointed president of Magdalen College, Oxford when Henry Clerke died in 1687, but was unsuccessful. James II appointed the man Smith had hoped would be his patron, Bishop Samuel Parker. Soon thereafter Smith was expelled from his fellowship, which he had held since 1667, because of his opposition to the attempt by James to pack the college with Roman Catholic fellows.

Nathaniel Vincent was also unsuccessful in his quest for preferment. A fellow and senior proctor at Clare College, Cambridge, Vincent failed to obtain a mandate for the mastership of Clare when the incumbent died in 1678. He remained a fellow there until his death in 1722. Henry Jenkes was fellow of Caius College, Cambridge for forty-four years. He was lecturer in Greek when he was elected to the Royal Society in 1674, and was to become lecturer in Hebrew four years later. Though he was proctor at Caius from 1686 to 1687, and dean and chaplain for many years, he never managed to rise any higher. He was close to another clerical virtuoso who never tried to rise. The Cambridge Platonist, Henry More, was widely respected for his learning, but he steadfastly refused all offers of preferment. More remained a fellow of Christ's College, Cambridge from his election in 1639 to his death in 1687.

A number of clerical virtuosi did not stay at Oxford or Cambridge beyond the completion of their M.A., but resigned their fellowships to take advantage of preferment elsewhere. Several clerical virtuosi resigned their fellowships to become private chaplains. William Wotton resigned his fellowship at St. John's College, Cambridge, to become chaplain to Daniel Finch, Earl of Nottingham. John Rogers left Oxford with his M.A. to become chaplain to George, Earl of Berkeley, who was himself a fellow of the Royal Society. Shortly after resigning his fellowship at Wadham College, Oxford in 1681, Thomas Pigot became domestic chaplain to James, earl of Ossory, at whose home he died in 1686.

James Arderne, Daniel Milles, William Payne, and William Holder were college dons who became London preachers. Arderne was perpetual curate of St. Botolph, Aldersgate, before rising to become dean of Chester in 1682. Milles resigned his college fellowship in 1658, to become rector of St. Olave, Hart Street. There are caustic references to his sermons in the diary of Thomas Pepys.[5] Payne left Magdalene College, Cambridge in 1675, for a rectorship in Kent, and there he would have stayed had he not received the rectorship of St. Mary Whitechapel in 1681. It was at St. Mary's that he gained a reputation for his preaching. William Holder resigned his fellowship at Pembroke Hall, Cambridge in 1642, for the hand of the sister of Sir Christopher Wren and the rectory of Bletchington, near Oxford. With the Restoration, he too managed to get himself to London by becoming sub-dean of the Chapel Royal.

Francis Potter, Joseph Glanvill, John Beale, and John Pell gave up their college fellowships to assume rural livings. Francis Potter, a close friend of John Aubrey, left Trinity College, Oxford, three years after receiving his B.D. in 1625, to succeed his father as rector of Kilmington in Wiltshire. He remained in this living until his death in 1678. Joseph Glanvill, apologist for the Royal Society and the new learning, left Lincoln College, Oxford upon the Restoration for a living near London in Essex. In 1666, he moved from the London area to assume the duties of rector of the Abbey Church in Bath. His tenure there lasted until his death in 1680. John Beale, scientific visionary and member of the Hartlib circle, was a fellow at King's College, Cambridge, until 1640. He resigned his fellowship in order to travel and at the Restoration, he received the rectory of Yeovil in Somerset, which he held to his death in 1682.

John Pell, the mathematician, was also a member of the Hartlib circle, and like Beale, retired to the country at the Restoration. He first became acquainted with Samuel Hartlib when he was a student at Trinity College, Cambridge in the

1620's. Pell taught school after receiving his M.A. in 1630, but sometime thereafter he was persuaded by Hartlib to migrate to London and actively pursue his studies in mathematics. Patronage for this endeavor was not forthcoming, so Pell emigrated to the Low Countries. He accepted a professorship in mathematics at Amsterdam in 1643, and moved to a similar chair at Breda in 1646. With the advent of the First Dutch War, Pell returned to England where he petitioned the council of state for a mathematical lectureship. The Cromwellian government offered him a diplomatic post instead.[6]

With the return of Charles II, Pell fell upon hard times. According to John Aubrey, Sheldon presented him with the parsonage of Fobbing in Essex in 1661, along with another "scurvy" Essex parsonage which the locals referred to as "kill-priest." The latter parsonage failed to work the effect that its name implied, for Pell did not die until 1685. The proceeds from both livings were never sufficient to relieve his poverty. Aubrey reported that he was thrown into prison for debt in 1680.[7]

Three clerical virtuosi resigned their college fellowships for secular positions. William Perrey migrated from Trinity College, Cambridge in 1679, to become librarian of the Royal Society. In 1681, he was made Gresham professor of music, a post he held until his death in 1713. John Mapletoft was not yet ordained when he left Trinity College, Cambridge to become a tutor in the family of the Earl of Northumberland. In 1660, he went abroad to study medicine and received his M.D. in 1667, whereupon he set up practice in London. In 1675, he assumed the chair in physic at Gresham College. His practice brought him into contact with Thomas Sydenham, the noted physician, and John Locke. Perhaps it was his contact with the latitudinarian divines who frequented the house of his relation, Thomas Firmin, which persuaded him to give up medicine and take holy orders in 1679. From 1686 to his death in 1721, Mapletoft was vicar of St. Lawrence Jewry, London. Robert Wood, another member of the Hartlib circle, was an intruded fellow of Lincoln College, Oxford during the 1650's. There is evidence that he left for Ireland after receiving his license to practice medicine in 1656, and that he was involved in the aborted scheme to form a college at Durham. Except for a brief stint as mathematical master of Christ's Hospital in London, for which he was expelled for his refusal to teach elementary courses in mathematics, he spent the bulk of the Restoration in Ireland, becoming at one point the chancellor of the diocese of Meath, an office that made him the most senior official in the diocese and the bishop's chief troubleshooter.[8] The botanist,

John Ray, scrupled to take the oaths at the Restoration and resigned his fellowship at Trinity College. At the invitation of his student, Francis Willughby, Ray retired to Willughby's Warwickshire estate where he tutored his two sons and continued his biological investigations.

Though the majority of the clerics who became fellows of the Royal Society were university trained, there were a few who did not receive their educations at Oxford or Cambridge. Isaac Vossius and Anthony Horneck, for example, received their educations abroad before emigrating to England. Vossius was one of the scholars at the court of Queen Christiana from 1649 to 1652. John Pearson invited him to England in 1670, largely in recognition of his scholarship on behalf of the authenticity of the Ignatian epistles. He was awarded a prebend of Windsor to support him financially and honored with the degree of D.C.L. from Oxford.[9] Anthony Horneck received his M.A. from Heidelberg and emigrated to England from Germany in 1661. In 1664, he became chaplain at Queen's College and vicar of All Saints', Oxford. He established his fame as a preacher and casuist when he migrated to London and became preacher at the Savoy in 1671, a position he held until his death in 1697.

Sir George Wheler was one of the few men of learning among the clerical virtuosi to benefit from training at one of the alternatives to an university education, namely, the Inns of Court. The Inns of Court in the seventeenth century were more than just a law school where, in the words of one contemporary, the "studious *Cato* may read *Littleton*, the more facetious and planetary spark Cook." According to this contemporary, the Inns of Court afforded "such latitude that students may have opportunity to buy Authours that treat of all subjects, and converse with proficients in all Sciences."[10] Though Wheler matriculated at Lincoln College, Oxford in 1667/8, no record exists of his receiving a B.A. In 1671, Wheler became a student of Middle Temple before travelling to France, Italy, and Greece in the middle 1670's.

The career of Edward Waterhouse, heraldic scholar and legal historian, supports the point that a reputation for learning was not indissolubly linked to formal university training. Though Waterhouse was admitted a fellow commoner at Emmanuel College, Cambridge in 1635, there is no record of his having ever received the B.A., much less the M.A. The only degree he was ever awarded was the LL.D., which he received by mandate in 1668, probably in recognition for his commentary upon Sir John Fortescue's *De Laudibus Legum Angliae* published in 1663.[11] Waterhouse enjoyed the conversation of university men during the years he resided in Oxford and studied at the Bodleian. His contribution to the Interregnum debate over reform of the universities, his *An Apology for Learning and Learned Men* (1653), indicates that he admired learned men, but he did not always succeed in emulating their productions.[12] His works show a notable lack of the very skill which the universities sought to engender, namely, the ability to organize, assess, and present the received wisdom on any subject. When the clerical virtuoso, Robert Wood, was once asked the use of an university education, he replied that it taught the art of "methodizing what others of excellent parts have invented."[13] Waterhouse's own endeavors often suffered from his inability to "methodize" his materials.

Learning and Wit

The third and last personal quality which Oldenburg affirmed belonged to many fellows of the Royal Society was wit. In its widest acceptation, the word, "wit," referred to the faculty of thinking and reasoning in general. It was a measure of intellectual ability, not intellectual achievement. Learning was a sign of wit, though wit, unlike learning, could not be cultivated. Wit itself was a gift and it might even be found in persons who had little or no learning. It was in recognition of this fact that Charles II bade the Society to accept men without regard for their station or their learning if they were reputed men of wit.[14]

The cleric, Thomas Baker, was one such man. He attended Oxford briefly in his youth, becoming a scholar at Wadham College, Oxford in 1645, and graduating with a B.A. in 1647. There is no record of his activities thereafter until he became vicar of Bishop's Nympton, Devon in 1660. Anthony à Wood reported that Baker lived there many years "in a most retired and studious condition."[15] None knew of his ability in mathematics until John Collins became acquainted with his genius and proposed that one of his works be published by the Society. *The Geometrical Key* was published with the Society's imprimatur in 1684, and its author was elected to the Society the very same year.[16] The book must have sold well, for two years later one Mr. Clavel, a bookseller, was encouraging the Society to print more of Baker's papers. The Society assured the bookseller that several members would purchase such a volume were it to be printed.[17]

Baker's career illustrates another attribute of wit which was both celebrated and distrusted in the seventeenth century. Baker's virtuosity in mathematics was not only indicative of the general acuity of his intellect, it also implied a particular cast of mind. Baker displayed his virtuosity, as did other wits, in the ingenuity of his productions. The mark of a wit was originality, resourcefulness, and cleverness. Originality was worthy of praise so long as it did not smack of singularity. Resourcefulness was worthy of emulation so long as it was not confused with disobedience. And cleverness was worthy of admiration so long as it was exercised on behalf of truth, and not applause. The seventeenth century regarded learning as an antidote for all three sins, inasmuch as learning had the virtue of teaching a wit judgment. Without the ability to judge the inventions of his wit, a person could easily mistake what were only flights of fancy for solid knowledge.

Conformity to tradition was the criterion of truth in matters of learning. The main task of a man of learning was to master the body of knowledge which met this test. If a learned man were a man of wit, he might be expected to add to this stock of knowledge, but his primary responsibility was the transmission of this stock to future generations. Since the man of learning defined truth in terms of obedience, conformity, and consensus, it is not surprising that he should have suspected the wit of being subversive to the very values it was his social function to uphold.

Apologists for the Royal Society were acutely aware of these suspicions and sought to alleviate them by stressing the number of learned men who were members of the Society.[18] Both Sprat and Glanvill argued that a society which looked favorably upon wit also honored learning.[19] They emphasized that the study of nature, like any other learning, served to remedy the faults to which wit was liable. They claimed that such study dampened the impulse toward pride and made a man more humble.[20] They contended that conformity to nature was an adequate test for the inventions of wit, and corporate research an adequate safeguard against singularity.[21]

There is a danger of anachronism in the attempt to single out the wits among the clerical virtuosi if the only criterion used for their selection is their inclusion in the *Dictionary of Scientific Biography*. To judge the work of the Royal Society and the contributions of its members solely by how far they advanced the march of modern science encourages the dismissal of whatever seems antithetical to modern sensibilities. Historians of science, for example, have only recently come to appreciate how integral to the thought world of the seventeenth century was an "aberration" like alchemy.

Care should also be taken not to impose upon their accomplishments the disciplinary categories of modern science. Barbara Shapiro has reminded historians recently that "seventeenth century intellectuals were far less bound by disciplinary boundaries than their modern counter-parts."[22] Knowledge in many fields had yet to become complex and many fields of endeavor had yet to crystallize into disciplines. Shapiro contends that the fact that history and natural history still share a name in common is a survival from a time when the boundary between the two was far more fluid. Shapiro observes that the seventeenth century identified history as "a true account of the facts—any facts—physical, biological, social, and/or 'historical'."[23] The century assumed that the methods used to ascertain the facts in "civil" history were the same as those used in natural history. This may help to explain in part why an ecclesiastical historian like John Pearson was active in the Society. The methods he used to defend the authenticity of the letters of St. Ignatius did not differ substantially from the methods John Ray used to determine the proper genus for a plant. Both had to conjecture, often from fragmentary evidence, what the copy text or proper genus had to be.

It was also not unusual for research projects in both history and natural history to overlap. When Sir George Wheler travelled through Greece from 1675 through 1676, he was as assiduous in collecting plants as he was in collecting coins, classical manuscripts, and antique marbles. Wheler's praise for his guide is indicative of this overlapping. In his narrative of the journey, Wheler described his guide as,

> an admirable Fellow for our purpose, following us always on foot; and if he saw any old Church or Building near the way, he would presently run to it, to see if he could find any Inscription, or other mark of Antiquity upon it: which if he did, he would immediately call out to us, Scrittione, Scrittione... If he saw any Flower or Plant that he thought was not common, he would not fail to gather it, and bring it to me; saving me often the labour of lighting from my Horse, to gather those I observed my self.[24]

Whatever our estimation of their wit, it is the case that some clerics had a greater reputation for wit than others and that their contemporary reputations often matched their subsequent reputations for brilliance. Among those generally reputed to be wits were John Wallis, Isaac Barrow, John Pell, Seth Ward, John Flamsteed, John Wilkins, and John Ray.

John Wallis, Savilian professor of geometry, was not a modest man. When he published a history of algebra in 1685, he proudly proclaimed that it was upon his shoulders that Newton had glimpsed the method of the calculus. Enumerating for future ages the stages by which the study of the infinitesimal had advanced, Wallis placed himself third in a line that began with Archimedes, continued through Bonaventura Cavalieri, and ended with Isaac Newton.[25] Historians have acknowledged that Wallis was correct in his estimation of the significance of his discoveries, though he may have underestimated the impact that other mathematical traditions had upon Newton's development.

When Wallis published his *Arithmetica Infinitorum* in 1656, it fundamentally altered the balance between the visual and the abstract in English mathematics.[26] Cavalieri's method of indivisibles was essentially a geometrical method for measuring areas under curves. Wallis applied the tools of algebraic analysis, tools perfected by continental mathematicians, to finding areas under curves. He not only duplicated Cavalieri's results, but he demonstrated that algebraic analysis could solve area problems which were beyond the means of the geometrical methods of his day. Wallis recalculated pi in the course of finding an arithmetical expression for the area $\pi/4$ of a quadrant of a circle whose algebraic formulation was $x^2 + y^2 = 1$. Through sheer dint of mathematical calculation and a shrewdness for inspired interpolations, he found a value for pi, which, though it could not be stated exactly, constituted a limit which could be approached to any degree desired.[27]

The success of Wallis's methods were also the source of their weakness. They had little else than utility to recommend them. Their accuracy could be confirmed in cases where simple measurement was possible, but their validity remained in doubt. Newton did nothing to improve them upon this score; that had to wait until the nineteenth century.[28] What he did as a result of reading Wallis's work in the winter of 1664-1665, was to make his methods more flexible, and thus more general, thereby extending their range.[29]

Isaac Barrow would be of minor interest to historians of science if his name were not connected to that of Isaac Newton. Barrow was a fellow at Trinity College, Cambridge, when Newton matriculated there in 1661. Though Newton did not have Barrow for his tutor, as is commonly supposed, he benefited from Barrow's patronage at crucial points in his career. When Barrow, the first

Lucasian professor of mathematics, resigned his position in 1669, he engineered Newton's appointment. In 1675, Barrow, now master of Trinity, helped Newton obtain a royal dispensation that allowed him to hold his fellowship at Trinity without taking orders.[30] Newton showed his gratitude by editing his patron's lectures on optics and geometry. Barrow presented Newton with an inscribed copy of the latter when they were published in 1670.[31] Newton also also helped Barrow with the *errata* to his editions of Euclid and Archimedes.[32]

From the references to Barrow in Newton's papers, there is little to suggest that Barrow had a profound impact upon the development of Newton's thought. Newton reported that Barrow was his examiner when he stood for a scholarship at Trinity in 1664. Newton remembered doing badly in the examination because he had spent most of his time beforehand reading Descartes when he should have been studying Euclid. Nevertheless, he won a scholarship.[33] Barrow was certainly acquainted with Newton's genius by late 1668, when he lent him Nicholas Mercator's *Logarithmo-technia*. John Collins had sent the book to Barrow, and Barrow appears to have shown it to Newton in the hope that he would publish his own discoveries before Mercator advanced very much further. The result was Newton's paper on the analysis of infinite series which Barrow forwarded to Collins.[34]

J. M. Child has ventured the theory that Newton first grasped the fundamental theorem of the calculus as a result of Barrow's lectures.[35] This theorem states that there is an inverse relation between finding the area of a curve and finding its slope. If Barrow suggested this in his lectures, he did so in a form that was purely geometrical. There is no evidence in his lectures of calcular notation, rules, and symbols nor is it likely that he used a private calculus to work out his solutions before demonstrating them publicly by means of geometry.[36]

Of the impact of Barrow's lectures, Newton would only say that they might have put him "upon considering the generation of figures by motion."[37] Kinematic analysis of geometric curves was hardly novel when Barrow lectured upon the subject in 1666. Both Evangelista Torricelli and Giles Persone de Roberval had analyzed curves by regarding them as the paths traced by points in motion.[38] Though Barrow improved upon their methods, his improvements were strictly geometrical. He considered algebra a useful logical tool, but he did not see it supplanting the methods of classical geometry. His lectures attest to his having read the leading mathematicians of his day, but they also show that he continued to approach their work from the standpoint of Greek mathematics.

Seventeenth century geometers, like Barrow, were initially attracted to the study of algebra from the mistaken belief that it represented the method which the Greeks had used to make their geometrical discoveries.[39] Barrow continued to hold this belief at a time when his contemporaries were resolving problems by this method which the Greeks had never broached. Barrow did not see, as Newton did, that an algebraic formulation could express a curve far more precisely than a geometrical figure could. Algebra for him remained a useful logical tool, not a branch of mathematics.[40]

John Pell is largely remembered for an equation he never invented. The "Pell Equation" was the product of Pierre de Fermat's investigations of number theory. Thanks to a mistaken attribution by Leonhard Euler in the eighteenth century, it has borne Pell's name whose only contribution was to write it down in his papers. According to Fermat's biographer, the equation became the "central bone of contention" between Fermat and Wallis in 1657.[41] In an attempt to interest English mathematicians in the intricacies of number theory, Fermat challenged them to give a general solution to the equation $Ax^2 + 1 = y^2$.

The equation states the curious fact that an integer when multiplied by a square will result in another square if the number one is added to it. Thus the product of 3 multiplied by 1^2 will produce a square if 1 is added to the product. Fermat proclaimed that this would hold true for any integer and he challenged the English to prove why this was so.

Pell's own interest lay in applied mathematics. A member of the Hartlib circle, Pell advocated the compilation of a natural history of mathematics. Another of his projects was the establishment of a mathematical 'public library' which would disseminate as well as collect mathematical materials. It was Pell's intention that the library maintain a circle of correspondents and that it create a register of teachers and practitioners in mathematics. Teachers and practitioners would be licensed by the issuance of testimonials "after examination to all sorts of practisers, as Pilots, Masters, Landmeters, Accomptants, &c. of their speculative ability and practical dexterity, that those that have occasion to use such men, be not abused by unable undertakers, to their exceeding damage."[42] These suggestions were published in his *Idea of Mathematics* (1638), upon which his fame in the seventeenth century largely rested. It was republished by John Dury in 1650 in the *The Reformed Librarie Keeper*. Theodore Haak sent a copy of the *Idea* to Mersenne, who circulated the work. Descartes even replied approvingly.[43]

In 1647, Pell published *Controversiae de vera Circuli mensura* which attacked the mathematician, C. S. Longomontanus, and elicited approval from Roberval, Hobbes, Cavendish, Cavalieri, Descartes, and others.[44] He is also credited with editing Thomas Brancker's 1668 translation of J. H. Rahn's *Teutsche Algebra oder algebraische Rechenkunst* (Zurich, 1659).[45] This work introduced the symbol for division and was one of the first to set out equations in three columns, two for identification and one for explanation.[46]

Seth Ward's fame as an astronomer rested upon his discovery of a popular method for calculating ephemerides that approximated the results obtained using Kepler's second law of planetary motion.[47]

Kepler's second law states that the "velocity of a planet varies with its distance from the sun in such a way that a line joining the planet with the sun sweeps out equal areas in equal times."[48] The problem with this law was that it was almost impossible to apply before the invention of the calculus. Astronomers had the geometrical tools to deal with circles. Ellipses were beyond them, for no easy way existed for plotting areas under curves.

Astronomers before Kepler had assumed that the motion of the planets was uniform and circular. Their methods for calculating planetary positions had been built upon the assumption that a line drawn from a planet to the center of its orbit would sweep out equal arcs in equal times. Once they knew the radius of a planet's orbit and the time it took to complete one orbit, they could use geometry and trigonometry to calculate its position in the skies for any time requested.

Ward applied the equant method to approximating planetary positions under Kepler's laws. His innovation permitted astronomers to calculate planetary positions using traditional trigonometric methods. Ward proposed that the sun be that point where the velocity of a planet would appear to an observer to vary according to Kepler's second law. He reasoned that there must also be another point where the velocity of the planet would appear to an observer not to vary at all. A line drawn from this point to the planet would sweep out equal arcs in equal times. Ward conjectured that this point coincided with the empty focus of a planet's elliptical orbit, namely, the focus opposite the sun.

When John Flamsteed was named first Royal Astronomer in 1675, he was entrusted with the task of making an accurate survey of the heavens that would fix the positions of the stars and the motions of the planets. Though he had toiled at this task for over forty years and often at his own expense, he was disappointed in the end that the tolerance of his measurements would not permit the calculation of longitudes, the purpose for which his post had been established. Nevertheless, Flamsteed was able to achieve an accuracy in his observations that was fifteen times greater than that attained by Tycho Brahe a century earlier. The implications of this were enormous for astronomy inasmuch as Brahe's figures had provided the empirical basis for every theory in astronomy from Kepler onward. Flamsteed made over 20,000 observations using a sextant and a micrometer telescope in his first sixteen years as Royal Astronomer. These observations were to form the basis of his 3,000 star "British Catalogue" upon which his reputation rests.[49]

Flamsteed's approach to astronomy was uncompromisingly empirical. Flamsteed discovered quite early that the moon experienced irregularities in its orbit for which Kepler's laws could not account.[50] In Flamsteed's opinion, the only hypothesis that had come close to describing the actual motion of the moon was one invented by Jeremiah Horrocks (1618-1641). Flamsteed improved it and had it published in Wallis's edition of Horrocks's opera in 1673.[51] Horrocks's theory was no more than a method for calculating the motion of the moon. It did not explain why the irregularities occurred; it merely described them. Flamsteed continued to be interested in the problem and published revisions of Horrocks's theory in light of further observations.[52]

John Wilkins probably did more than anyone in England in his period to inculcate interest in natural philosophy.[53] He publicized the advances in science in a series of books that were written in the vernacular for the general public in the 1640's. As warden of Wadham College, Oxford in the 1650's, he made the college a mecca for natural philosophy. In the 1660's, he was instrumental in the formation of the Royal Society of London, for which he acted as secretary, facilitating its activities and overseeing its administration until his death in 1672.

Except for his *An Essay Towards a Real Character And a Philosophical Language*, he published the bulk of his scientific works in the 1640's.[54] His first work, *The Discovery of a World in the Moone. Or, A Discourse Tending, To Prove, that 'tis Probable there may be another habitable World in that Planet* was published in 1638.[55] In the *Discovery*, Wilkins argued that the moon was not a luminescent body, but a world like earth. He claimed that his hypothesis conformed with the observable phenomena and provided the most simple explanation for what could be seen through a telescope.[56] Two years later Wilkins published *A Discourse concerning A New Planet. Tending to prove, That 'tis probable our Earth is one of the Planets* (1640), which broached for the

English speaking public the cosmological theories of Copernicus, Kepler, and Galileo.[57]

In his next book, *Mercury, or the Secret and Swift Messenger: Showing, How a Man may with Privacy and Speed communicate his Thoughts to a Friend at any distance*, published in 1641, Wilkins addressed himself to the problem of communications.[58] He not only discussed such practical issues as how to protect the privacy of communications using ciphers and code, he also discussed the nature of communication and the relation of language to the world.[59] This involved an assessment of the natural language theories of Robert Fludd and others.

Both Wilkins and the natural language theorists agreed that language was flawed because it did not reflect the nature of things. Both assumed that this flaw entered language as a result of Babel. The natural language theorists conjectured that language would once again mirror nature if Adam's language could be restored. Once its vocabulary and grammar were known, there would no longer be any difference between what was said and the means used to say it.

Wilkins rejected their approach. His approach was 'analytic.' Rather than search for the principles of Adamic language in the hope of gaining knowledge of the world, Wilkins assumed what his opponents sought. Wilkins assumed that man already possessed some knowledge of the world and in his *Essay Towards a Real Character and a Philosophical Language* (1668), he tried to construct a language that would be symbolically equivalent to the world as man knew it. A language so constructed would continue to grow as man's knowledge of the world increased so that over time this language would approximate the original language which Adam spoke.

The last scientific work Wilkins published before being named warden of Wadham College, Oxford, was *Mathematical Magick*.[60] Published in 1648, the title indicates that Wilkins's subject was not the theoretical advances of geometers like Barrow or algebraists like Wallis, but the "magic" that resulted from the clever application of mechanical devices whose effects could only be understood by an appeal to mathematics. He hoped the clever application of mathematics would prove of benefit to gentlemen and artisans.

Being a good publicist, he was more interested in provoking wonder than explaining to his readers the mathematics behind these marvels. He discussed simple tools like the balance, the lever, the wheel, the pulley, the wedge, and the screw, and he included illustrations which became increasingly more graphic the more he exercised his ingenuity in inventing incredible applications. Wilkins, for example, made the claim at one point that a combination of these tools could produce "infinite strength" so as to "pull up any oak by the roots with a hair, lift it up with a straw, or blow it up with one's breath."[61] Wilkins also discussed such things as flying machines, perpetual motion, statues that moved and talked, artificial spiders, imitation bird calls, prairie schooners, submarines, and the Archimedean screw.[62]

Wilkins had a chance to demonstrate some of the curiosities discussed in *Mathematical Magick* when he was appointed warden of Wadham College, Oxford in 1648. On display in the garden at Wadham were such marvels as a hollow statue that spoke, beehives from which the honey could be removed without destroying the colony, waterworks, and plants both useful and curious.[63] As warden, Wilkins so improved the college's finances and reputation that it became universally known as a refuge for learning and discussion. Wilkins was influential in founding the Oxford Experimental Club which met at Wadham during his tenure and attracted such notable virtuosi as Walter Pope, Laurence Rooke, William Neile, Thomas Sprat, Robert Boyle, John Locke, William Petty, Seth Ward, and Christopher Wren.[64] Wilkins remained warden of Wadham until 1659, when he migrated to Cambridge to become master of Trinity College. Though his tenure there was brief, he made contacts in Cambridge that proved of use when he organized the Royal Society in 1660.

Wilkins would have been the Society's first president, had his Cromwellian connections not made it impolitic in the climate of the Restoration. Wilkins was elected to the post of secretary instead and he attended nearly every meeting of the Society until his elevation to the diocese of Chester in 1668, forced him to reduce his activity on behalf of the Society. Wilkins had a knack for eliciting cooperation and managing organizations. He concerned himself with all aspects of the Society's affairs from designing experiments for the visits of royalty to suggesting that Hooke be nominated for the office of curator. He also supervised Sprat's authorship of the *History of the Royal Society*. His supervision was so close that Sprat commented in the text that it was "onely my hand that goes, the substance and direction come from one of them," namely, either Wilkins or Oldenburg.[65]

It was during this period that Wilkins published *An Essay Towards a Real Character and a Philosophical Language* in which he returned to the subject he had first touched upon in the *Mercury*, back in 1641. His scheme to construct an universal language which would avoid the faults which particular languages were liable was admired for its ingenuity, but the actual result fell far short of expectations. It met with embarrassment when it was published in 1668, and was quietly forgotten.

John Ray is primarily remembered for his work in botany. His catalogue of British plants remained foundational for English botanists until well into the nineteenth century and his *Methodus Plantarum Nova* (1682) was among the first books in botany devoted to the problem of classification.[66]

The problem facing naturalists like Ray in the mid-seventeenth century was how to cope with the sheer profligacy of nature. This problem was particularly acute in botany. A. R. Hall reports that botanists had described some six thousand distinct plants by 1600, and were to triple this number by the end of the century.[67] Their success in collecting specimens threatened to overwhelm the primitive systems of classification inherited from the ancients. Ancient authors had listed plants alphabetically or grouped them artificially under categories like size or use. Ray had used an alphabetical listing in his first botanical work, the anonymous *Catalogus Plantarum Circa Cantabrigiam nascentium* (1660), and he never totally freed himself from the influence of Theophrastus's tripartite division of plants by height into shrubs, trees, and herbs.[68] Though such systems worked well enough when the number of plants to be classified was small, (Theophrastus had dealt with approximately four hundred fifty plants), it proved unmanageable when the number reached the thousands.[69]

One strategy which botanists tried was to catalogue plants according to their form or structure. Morphological analysis promised to produce more "natural" groupings than the "artificial" systems of the ancients, provided some consensus could be reached concerning which anatomical characteristics were essential, and thus criterial, and which merely accidental. Such a consensus eluded botanists in the seventeenth century. Some like Andrea Cesalpino divided plants according to the variations exhibited by one trait, in his case the variations exhibited by fruit or seed. Unfortunately, reliance upon a single trait could produce groupings no less artificial than those devised by the ancients.

Ray preferred to group plants by the principle of family resemblance. He refused to exclude a species from membership in a genus because it failed to exhibit some crucial characteristic of flower, fruit, or seed. He would include it if he could show that the species in question shared other characteristics with the genus it resembled. Of course, these characteristics had to be invariable or the identification would be faulty. Similarities based on size, scent, taste, or habit were invalid because they varied not only among species within a genus, but also among individuals within a species. Ray was among the first to recognize that the tendency of seeds to germinate with either one or two leaves was an invariable characteristic, and his discovery figured prominently in his classificatory scheme. Using the rules outlined above, Ray compiled the work upon which his botanical fame rests, his *Historia Plantarum* (1686-1704), a three volume encyclopedia of all the plants then known in Europe.[70]

Though botany remained central to his interests, Ray also engaged in zoological research. The occasion for this research was the death of Francis Willughby, Ray's patron, friend, and collaborator. Ray had originally intended his *Historia Plantarum* to be his contribution to the "systema naturae" upon which both men had been working since their days at Cambridge before the Restoration.

Out of respect for his friend, Ray completed the catalogues his friend had been compiling on fishes and on birds, and published them both under Willughby's name in 1676 and 1686.[71] He then went on to catalogue mammals, reptiles, and insects in an effort to complete the project. Ray brought to these investigations the same attention to taxonomy which had characterized his descriptions of plants.

Not only was Ray interested in the language of nature, he also shared an interest in the nature of language. He contributed the botanical tables for John Wilkins's *Essay Towards a Real Character and a Philosophical Language* (1668), and it was his dissatisfaction with its "author's prescribed system" that led him to compose his own method of classification in his *Methodus Plantarum Nova* of 1682. His work on Wilkins's *Essay* was not his only incursion into philology. In 1670, he published a collection of English proverbs.[72] His concern with method is evident here, for his title proclaimed that the proverbs contained therein were "Digested into a convenient Method for the speedy finding any one upon occasion." Philologists continue to find his work invaluable as a source for dialect and folklore. Four years later, Ray published a collection of English words in which he advocated reforms in English spelling.[73]

Ray is best known, however, for two works of theology which were published late in life though they had their origins in the period before Ray left Cambridge in 1662. Ray claimed that his *Miscellaneous Discourses* (or *Three Physico-Theological Discourses*, as the work was styled in subsequent editions) was the outgrowth of a sermon he had delivered at St. Mary's Church in

Cambridge "30 years agoe."[74] Likewise, Ray had also preached the germ of his *Wisdom of God Manifested in the Works of the Creation* at morning divinity exercises in Trinity College Chapel.[75] The latter, composed at the culmination of Ray's scientific career, constituted for many in the eighteenth century their first introduction to natural philosophy. Ray's biographer, C.E. Raven, claims that the work "formed the basis of Derham's Boyle Lectures in 1711-12; it supplied the background for the thought of Gilbert White and, indeed, for naturalists of three generations; it was imitated and extensively plagiarized by Paley in his famous *Natural Theology*; and more than any other single book it initiated the true adventure of modern science, and is the ancestor of *The Origin of Species* or of *L'Evolution Creatice*."[76] Using a wealth of illustrations not only from botany and zoology, but also from astronomy and geology, Ray argued that creation could not have been a matter of chance nor was the world maintained by mechanical laws alone. Nature was so infused with purpose that the world would be inexplicable if God did not exist. The study of nature was the study of God's wisdom. In his *Miscellaneous Discourses*, Ray argued that it was possible to construct hypotheses for the Deluge and Creation that would not conflict with the account in Scripture, and he attempted to explain why fossils, "Sea-Shells found in the Earth" and "Subterra-neous Trees" had not survived until the present day despite the widespread belief that God had saved every species in the ark.

Though the accomplishments of Wallis, Barrow, Ward, Flamsteed, Wilkins, and Ray are still honored today, their names do not exhaust the number who enjoyed notoriety for their wit.

St. George Ashe was typical of the virtuosi of his period in the breadth of his interests. Between 1683 and 1687, he read at least twenty-eight papers before the Dublin Philosophical Society on subjects that ranged from the squaring of the circle to experiments in freezing.[77] The quality of his work was similarly varied. The paper Ashe gave on the squaring of the circle indicates that he was well abreast of the mathematical authors of his period.[78] On the other hand, his experiments in freezing showed a singular lack of rigor. Ashe merely subjected a number of substances like pumpwater, saltwater, ale, milk, claret, aniseed, water, urine and syrup of gillyflowers to the effects of the cold to see what would happen.[79]

Ashe did some of his best work in astronomy and meteorology. It was through his urging that Trinity College, Dublin purchased a moderately sized telescope. Ashe wrote John Flamsteed to ask him about the most suitable instrument to purchase.[80] In July of 1684, Ashe helped direct the Dublin effort of a joint project to observe a scheduled solar eclipse simultaneously in London, Oxford, and Ireland.[81] Another subject for which he expended considerable energy was the weather.[82] In the years 1685 and 1686, Ashe kept eight monthly diaries on the weather, noting barometric pressure, the temperature, the strength and direction of the wind, and the condition of the skies. Ashe recognized the importance of keeping long term records of the weather and comparing them with the statistics kept in other locations. He also recognized the difficulties inherent in the project inasmuch as instruments like thermometers lacked a common scale for measuring variations. He even allowed himself to speculate on what made Dublin weather so unwholesome. He concluded that it was the nearby Wicklow Mountains that collected clouds which brought rain, and which kept "the air and smoak pent up stagnant over us, preventing the motion and agitation which is necessary to preserve it pure and defaecated."[83] On the basis of his observations, Ashe suggested that industries which polluted the air should be located on the east coast away from Dublin.

Robert Wood, the mathematician, was a pupil of William Oughtred whose work, the *Clavis mathematica* (1631), was one of the most popular textbooks of the period.[84] Oughtred acknowledged Wood's help in preparing an English translation of the work in 1647, and he called upon Wood again in 1651, to help Wallis and Ward prepare another to be published in Oxford in 1652.[85]

Like John Pell, Wood had an interest in applied mathematics. In March of 1656, Wood forwarded Hartlib a scheme addressed to Cromwell and his advisers for the decimalization of the English currency. He titled his manuscript, "Ten to One."[86] Wood was realistic enough to recognize the inherent difficulties of his scheme. Though the current effort to mechanize the production of English coinage made the introduction of his reform opportune, the new coinage would have to compete against the old. Wood believed that its greater ease in reckoning would serve to drive the old money out of circulation. He also believed that this reform stood a greater chance of success than the decimalization of weights and measures since the value of money was not set by custom nor did it vary in different parts of the nation.[87]

Wood continued to lobby for his scheme after he left Oxford for Ireland in 1656. His interest in the scheme led him to speculate about the value of money in general. He formulated some of these thoughts when he engaged in a debate in 1658 and 1659 with one William Potter about the intrinsic value of bullion.

Potter argued that gold and silver were essentially like commodities. They had no intrinsic value beyond that accorded them by market forces. Wood disagreed. He felt that gold would still be valued even if its connection with currency were severed. Its aesthetic qualities gave it intrinsic value which made it desirable in its own right. He criticized Potter's scheme of minting coins from baser metals because the amount of metal they would require to match the value of the gold they would replace would make them too heavy for commercial use. The debate floundered on the inability of each side adequately to define the term, intrinsic value, or assign criteria for fixing a particular value to base or precious metals.[88] Wood's friend, William Petty, later proposed that the common measure for determining value was work. If the same amount of work was used to produce one bar of silver and five bars of iron, then the two quantities were equivalent in their intrinsic value.[89]

Wood also shared Hartlib's interest in agriculture. Upon his arrival in Ireland in 1657, Wood commenced experimenting with "clover husbandry" to enrich fallow fields, and he reported to Hartlib his observations on butter and cheese and the causes of loughs, the Irish term for a bay or inlet of the sea.[90]

Like Robert Wood, John Beale too was a member of the Hartlib circle. He was one of a number of Puritan authors who were interested in improving agriculture by collecting information about current practices, subjecting them to testing, and experimenting with new methods where older practices were deficient. Beale's interest was orchards. In 1658, he published *Herefordshire Orchards, A Pattern for all England* which discussed the propagation and cultivation of fruit trees, along with suggestions for improvement of cider production. Beale recognized that soil and climate were critical factors in determining how well a particular species of tree would grow, and he backed his opinions up with systematic observations. It was his conclusion, for example, that apples which had suffered the ravages of nature, that is, apples that were "more wrethed or wrinkled, or spotted with warts, moles, or freckles, or of a more russet, or yellow colour" made the best cider.[91]

His interest in these subjects continued into the Restoration. Correspondence from Beale concerning cider was read at a meeting of the Society on 23 March 1663/4. On 9 April 1673, another letter was read concerning cider grafts, and yet another one was read a month later which contained information on the grafting of roots to speed growth in an orchard, the mixing and alternating of vegetables, and the methods by which a tree could be made to bear many kinds of fruits such as apples, pears, nuts, grapes, plums, and cherries.[92]

Motives For Joining the Royal Society

To assert that the clerical virtuosi were men of birth, learning, and wit does not explain why this particular group of clerics became fellows of the Royal Society. If birth, learning, or wit made a cleric worthy of consideration, what factors led to his consideration or election? Robert G. Frank has suggested that the founding of the Royal Society represented an attempt on the part of university men "to continue, indeed to perpetuate, the kind of active association concerning learned topics which they found in their universities."[93]

Frank's suggestion, if true, would explain why a large proportion of the clerical virtuosi were university trained. Every cleric in this group of fifty-three spent some time at university either in England or abroad. Forty-eight out of the fifty-three pursued the seven year course of study to receive the M.A. As a result of their education, at least forty-eight clerics could boast some competence in science at the time of their election. This conclusion is based upon the fact that forty percent of the M.A. course of study consisted in geometry, astronomy, and natural philosophy, whereas the four year B.A. course of study dealt exclusively with logic, grammar, rhetoric, and Greek.[94] It is Frank's contention that the man who pursued his education to the M.A. or beyond, was "much more able to define important problems for himself, to carry out the work necessary to solve them, and to have the inclination and the confidence to present his conclusions to his peers for evaluation."[95] At least half of the clerical virtuosi proceeded beyond the M.A. Excluding creations, twenty-six clerics were awarded the D.D., two the B.D., and three the M.D.

University training, however, cannot be the only factor governing election to the Society. There were a number of individuals with university degrees and scientific interests who did not become fellows of the Royal Society. In his study of the Oxford scientific community from 1640 to 1675, Frank found one hundred ten individuals who showed some interest in science.[96] He identified twenty-seven of these as clerics. Of the twenty-seven, only eleven eventually became members of the Royal Society. What motives or circumstances did these eleven clerics share that made their subsequent election to the Royal Society likely?

Frank's research suggests that these eleven were distinguished by the number of close scientific friendships they formed with one another during their years at Oxford. Only one of the eleven did not form at least one close association with another cleric from the eleven. Wallis, Ward, Wilkins, and Bathurst were the objects of most of these friendships. Each had as many as five close friends among the eleven, and they shared many friends in common.

This web of friendships was notably lacking among the sixteen Oxford clerics who did not become fellows of the Royal Society. Only one of the sixteen was friendly with another cleric from the sixteen. None of the sixteen was a leading figure in Oxford science and only six forged close friendships with more than one of the leading Oxford scientists. The eleven who became members of the Royal Society, by contrast, had three of the leading Oxford scientists in their ranks, and close contacts with other leading Oxford scientists. On the whole, the eleven who became fellows of the Royal Society had more ties among the scientific community of Oxford than the sixteen who never became fellows of the Royal Society.

Frank has observed that friendship in the Oxford scientific community often formed among those the same age and along collegial lines.[97] This appears to be the case for the eleven who subsequently joined the Royal Society. These eleven maintained some fifteen friendships between themselves. Seven of these friendships were between clerics who were nearly the same age, and five were between clerics who were residing in the same college.

There were only two friendships out of the fifteen which were not between clerics of the same age cohort or college. In both cases, the two friendships were between clerics who had some association with the Oxford Experimental Club which met informally at Wadham College throughout the 1650's. Indeed, participation in the Oxford Experimental Club was the greatest common denominator linking all eleven clerics. Only two of the eleven clerics who eventually became fellows of the Royal Society had no association with the earlier Oxford club. Of the nine who were members of the Oxford Experimental Club, four were subsequently proposed for membership in the Royal Society by former participants in the Club, and only one of the nine was nominated to the Society by a person who had had no association with the Club.

It would appear, therefore, that friendship, particularly extensive ties of friendship, within the Oxford scientific community raised a cleric's chances of being proposed for membership in the Royal Society. Does this hold for clerics who were not members of the Oxford scientific community, but who resided in Oxford during the period under study and who were subsequently elected to the Royal Society? The answer is unclear. It is dangerous to ascribe friendship when the only available evidence is circumstantial, but it does appear that the eight clerical virtuosi whom Frank does not mention, (presumably because he found no evidence of either scientific interest or activity during their residence at Oxford), were at least known to the Oxford scientific community.

Edward Bernard was clearly a member of the Oxford scientific community before the end of 1675, and he was the only one of the eight who should have been included in Frank's inventory of the Oxford scientific community. Bernard matriculated at St. John's College, Oxford in 1655, and was made a fellow in 1658. He studied mathematics under John Wallis and it is a testimony to their friendship that Wallis put his son in his care when Bernard made a trip to Leiden in 1669. Bernard was carrying on a correspondence with Henry Oldenburg as early as 1670. In 1672, Bernard complained to Oldenburg how "the want of an Observatory damps Astronomique studies here" in Oxford.[98] In 1673, he succeeded Wren as Savilian professor of astronomy and was elected a fellow of the Royal Society upon the recommendation of Henry Oldenburg. Bernard, a noted linguist, proffered his knowledge of ancient sources to the effort to recover the mathematics of the ancients.[99]

Except for Bernard, there is only indirect evidence for the possible ties of friendship that existed between the eight clerics whom Frank did not list as participants in the Oxford scientific community and the one hundred ten individuals whom Frank did. Six of the eight lived in colleges where there were scientific coteries. Gilbert Sheldon was briefly warden of All Souls at the Restoration when Christopher Wren, George Castle, Thomas Millington, and John Mayow were residing within its walls. Sheldon himself had attended Trinity College, Oxford in 1620's, when the mathematicians, Thomas Allen and Henry Gellibrand, were fellows there. George Morley was dean of Christ Church briefly at the Restoration when the college boasted such luminaries as Thomas Willis, John Ward, Richard Lower, Robert Hooke, Nathaniel Hodges, and John Locke. John Dolben also had ties with Christ Church both before and after the Interregnum. He was among the group of strong Royalist divines who met at the house of Thomas Willis during the 1650's. Richard Allestree was one of the divines with whom Dolben was close. Allestree's interest in chemistry was strong enough for Frank to list him among the Oxford scientific community. The poet and cleric,

Samuel Woodford, received his B.A. from Wadham College in 1657, when Wilkins was warden. Wilkins subsequently proposed him for membership in the Royal Society in 1664. Thomas Baker is recorded as having received a B.A. from Wadham College in 1647, but it is not known whether he continued in residence long enough to have witnessed the flowering of interest in scientific subjects that occurred with the arrival of John Wilkins as warden in 1651.

Edward Wetenhall was probably in Oxford too short a time to have developed many ties with the Oxford scientific community. Wetenhall stayed only a year at Lincoln College, before leaving Oxford with his M.A. in 1661. He had migrated from Trinity College, Cambridge, the home of Cambridge science in the 1650's and 1660's, and as a fellow of Trinity under the mastership of Wilkins, Wetenhall probably knew John Ray, John Mapletoft, Thomas Gale, and Isaac Barrow, who were also fellows at the college during the period of his residence.

Anthony Horneck is the only cleric among the eight whose possible connection to the Oxford scientific community was not collegial. During the 1660's, Horneck was chaplain at Queen's and vicar of All Saints'. In 1665, he became chaplain to Christopher Monck, the second Duke of Albemarle. Seth Ward, who was friendly with George Monck, the second Duke's father, had tutored the son in mathematics when Ward was bishop of Exeter between 1662 and 1667. Whether it was Ward who recommended Horneck for the job of chaplain is not known, but it is the case that Ward sponsored Horneck for membership in the Royal Society.[100]

Table 2.1 contains the names of some clerics who were active in science during the course of their residence at Oxford, but who are not included in Frank's list of one hundred ten individuals because their residence at Oxford fell outside the time frame of his study. It appears that their careers were also marked by the pattern of friendships we have already noted. The friendships they made at Oxford they were able to continue as fellows of the Royal Society. Francis Potter, for example, was reputed to have been one of the first to have experimented with the transfusion of blood. Potter left Oxford before 1640, the starting point of Frank's study, but he remained friendly with John Aubrey, a Trinity man like himself. They exchanged letters during the 1650's, upon such topics as astronomy, logarithms, and mathematical instruments, and both subsequently became fellows of the Royal Society upon the Restoration.[101]

Also absent from Frank's list is the cleric, Thomas Pigot, who was active in Oxford scientific circles after 1675, the cutoff point of Frank's study. He was a fellow of Wadham College when he was elected to the Royal Society in 1679. In 1682, Pigot was one of the founders of the Oxford Philosophical Society whose founders included John Wallis, Ralph Bathurst, Edward Bernard, and Thomas Smith, all clerical virtuosi.[102] He appears to have been friendly with John Wallis, for Wallis forwarded one of his letters for publication in the *Philosophical Transactions*. The letter dealt with an earthquake that was felt in Oxford in 1683.[103]

The cleric, Thomas Smith, also missed inclusion in Frank's list. The first recorded instance of his interest in science occurred in 1675, when he published a sermon, dedicated to Robert Boyle, which used the new learning to defend the Christian religion.[104] John Wallis, with whom Smith was later friendly, mentioned his name as early as 1668 in a letter to Oldenburg, but only in connection with an inquiry about those in Oxford who were "skilled in Arabick."[105] In 1668, Smith went to Greece for three years as chaplain to Sir Daniel Harvey, ambassador to Constantinople. He presented the Bodleian with a number of Greek manuscripts upon his return, but he declined the entreaties of Bishop Pearson and Bishop Fell to make another trip to search for more. He was elected to the Royal Society in 1677, and from 1678 to 1679, he was chaplain to its president, Sir Joseph Williamson, a Queen's College man like himself. Smith was one of the founding members of the Oxford Philosophical Society in 1682, and a paper he read at one of its meetings was later published in the *Philosophical Transactions*.[106] After the Glorious Revolution made Smith a non-juror, he became librarian of the Cottonian library, and he carried on an extensive correspondence with men of learning all over Europe. He was a friend of Samuel Pepys, John Wallis, and Thomas Hearne.

It seems fair to conclude that the friendships a cleric made during his university years increased his chances of becoming a fellow of the Royal Society thereafter. Proximity to London pales next to friendship as a factor determining membership. Table 2.2 demonstrates that only half of the clerical virtuosi were residing in London at the time of their election. The proportion is the same for those eleven mentioned earlier who were members of the Oxford scientific community between 1640 and 1675, and who subsequently became members of the Royal Society. Yet residence in London exercised some influence, for when these eleven are compared with their sixteen clerical colleagues who were never elected to the Royal Society, the proportion of the latter who spent some part of their

TABLE 2.2

THE CLERICAL VIRTUOSI AND THE STATUS OF THEIR CAREERS AT THE TIME OF THEIR ELECTION TO THE ROYAL SOCIETY

Name	Date of Election to Royal Society	Status
Bishops		
Dolben	3/1665	bishop of Rochester
Henchman	3/1665	bishop of London
Laney	11/1666	bishop of Lincoln
Morley	11/1666	bishop of Winchester
Sheldon	3/1665	archbishop of Canterbury
Stearne	3/1665	archbishop of York
Wetenhall	7/1683	bishop of Cork and Ross
Cathedral Clergy		
Cotton	4/1663	archdeacon of Cornwall and canon of Exeter
Wilkins	11/1660	dean of Ripon (residing in London)
Williams	2/1664	vicar-choral of Bangor
London Clergy		
Arderne	6/1668	perpetual curate, St. Botolph Aldersgate
Milles	1/1675	rector, St. Olave, Hart Street
Payne	11/1681	rector, St. Mary Whitechapel
Tillotson	1/1672	lecturer, St. Lawrence Jewry; preacher, Lincoln's Inn
Ward	12/1661	vicar, St. Lawrence Jewry
Oxford or Cambridge Pastorates		
Horneck	1/1669	vicar, All Saints', Oxford

TABLE 2.2—*Continued*

Name	Date of Election to Royal Society	Status
Rural Clergy		
Baker	11/1684	vicar, Bishop's Nympton, Devon
Beale	8/1663	rector, Sock, Somerset, & rector of Yeovil, Somerset
Glanvill	12/1664	rector, Wimbish, Essex
Holder	1/1661	minister, Mere Hospital, Lincoln
Hotham	1/1668	minister to Bermuda
Pell	8/1662	rector, Fobbing, Essex
Potter	11/1663	rector, Kilmington, Wiltshire
Chaplains		
Rogers	4/1681	George, Earl of Berkeley
University Chairs and Lectureships		
Ashe	2/1686	Donegal lecturer & professor of mathematics, Trinity College, Dublin
Barrow	9/1662	Regius professor of Greek, Cambridge
Bernard	4/1673	Savilian professor of astronomy, Oxford
Castell	4/1674	professor of Arabic, Cambridge
Jenkes	11/1674	Greek lecturer, Cambridge
Pearson	3/1667	Lady Margaret professor of divinity, Cambridge
Wallis	3/1661	Savilian professor of geometry, Oxford

TABLE 2.2—*Continued*

Name	Date of Election to Royal Society	Status
College Heads		
Pearson	3/1667	Master, Trinity College, Cambridge
College Fellowships		
Bathurst	8/1663	fellow, Trinity College, Oxford
Clerke	10/1667	fellow, Magdalen College, Oxford
Jenkes	11/1674	fellow, Caius College, Cambridge
More	5/1664	fellow, Christ's College, Cambridge
Perrey	11/1678	fellow, Trinity College, Cambridge
Pigot	12/1679	fellow, Wadham College, Oxford
Smith	12/1677	fellow, Magdalen College, Oxford
Sprat	4/1663	fellow, Wadham College, Oxford (residing in London)
Vincent	11/1683	fellow, Clare College, Cambridge
Woodroffe	4/1668	tutor, Christ Church, Oxford
Wotton	12/1687	fellow, St. John's College, Cambridge
Gresham College		
Barrow	9/1662	Gresham professor of geometry
Jenkes	11/1674	Gresham professor of rhetoric

TABLE 2.2—*Continued*

Name	Date of Election to Royal Society	Status
Gresham College – ***Continued***		
Mapletoft	2/1676	Gresham professor of physic
Barrow	9/1662	Gresham professor of geometry
Jenkes	11/1674	Gresham professor of rhetoric
Mapletoft	2/1676	Gresham professor of physic
Inns of Court		
Woodford	11/1664	student, Inner Temple
Flamsteed	2/1677	Royal astronomer
Gale	12/1677	high master, St. Paul's School
Wood	4/1681	mathematical master, Christ's Church Hospital
Miscellaneous		
Burnet	3/1664	Scotland (in transit to Rotterdam)
Ray	11/1667	residing in Warwickshire
Vossius	4/1664	Leyden (?)
Waterhouse	7/1663	residing in Sion College, London

SOURCES: *Dictionary of National Biography*, 22 vols. (Oxford: Oxford University Press, 1885-1900); Joseph Foster, *Alumni Oxonienses*, 4 vols. (Oxford: Parker & Co., 1891-92); Robert G. Frank, *Harvey and the Oxford Physiologists* (Berkeley: University of California Press, 1980); A. Rupert Hall and Marie Boas Hall, trans. and ed.,*The Correspondence of Henry Oldenburg*, 13 vols. (Madison: vols. 1-9, The University of Wisconsin Press, 1965-75; London: vols. 10-11, Mansell, 1975-77; London: vol. 13, Taylor & Francis, 1986); K. Theodore Hoppen, *The Common Scientist of the Seventeenth Century* (Charlottesville: University Press of Virginia, 1970); John Le Neve, *Fasti Ecclesiae Anglicanae, 1541-1857*, 4 vols. (London: The Athlone Press for The Institute of Historical Research, 1969-75); A. G. Matthews, *Calamy Revised* (Oxford: Clarendon Press, 1934) and Idem, *Walker Revised* (Oxford: Clarendon Press, 1948); John Venn, *Biographical History of Gonville and Caius College*, 4 vols. (Cambridge: Cambridge University Press, 1897-1912), vol. 1: *Admissions 1349-1713*; John Venn, and J. A. Venn, *Alumni Cantabrigiensis*, 4 vols. (Cambridge: Cambridge University Press, 1922-54); Charles Webster, *The Great Instauration* (New York: Holmes & Meier Publishers, 1976).

lives in London is considerably smaller. As the next chapter will disclose, proximity to London was less important in determining whether a cleric would join the Royal Society than whether he would be active once he joined.

The ascription of motives from the facts of behavior is a risky enterprise in the absence of testimony. Unfortunately, none of the clerics examined chose to say why he joined the Society. One can speculate upon their motives, but such speculation can seriously misrepresent their actual reasons for joining. It seems reasonable to conclude from the pattern of their friendships that the desire to perpetuate associations forged at university was one of their motives for joining the Society, but the relative strength of this motive in the decision of any one cleric was a matter of individual circumstances. Given the nature of the Society and its reputation for exclusivity, a desire for the prestige of membership could have been an equally powerful motive in many cases. The Royal Society was not a group that one could decide to join. It was selective. One had to be invited. Invitations were not issued merely because a prospective fellow expressed an interest in natural philosophy. There is no evidence that some ever did, and yet they were not excluded for this reason. Those who were admitted usually listed themselves as F.R.S. along with their degrees on the title pages of their publications. This indicates that they were proud of their membership, and if the defenses issued on behalf of the Society are to be believed, it attested to their standing in society, to the strength of their wit, to the breadth of their learning, and to the sagacity of their judgment. The next chapter will examine the record of their activity once they accepted the Society's invitation to join.

NOTES

1 Charles Webster, *The Great Instauration: Science, Medicine and Reform, 1626-1650* (New York: Holmes & Meier Publishers, 1976), p. 142.

2 Joseph Foster, *Alumni Oxonienses: The Members of the University of Oxford, 1500-1714: Their Parentage, Birthplace and Year of Birth, with a Record of their Degrees. Being The Matriculation Register of the University, Alphabetically Arranged, Revised, and Annotated*, 4 vols. (Oxford: Parker & Co., 1891-92); John Venn and J. A. Venn, *Alumni Cantabrigiensis: A Biographical List of All Known Students, Graduates and Holders of Office at the University of Cambridge from the Earliest Times to 1900*, 4 vols. (Cambridge: Cambridge University Press, 1922-54).

3 Henry Oldenburg, *The Correspondence of Henry Oldenburg*, trans. and ed. by A. Rupert Hall and Marie Boas Hall, 13 vols. (Madison: vols. 1-9, The University of Wisconsin Press, 1965-75; London: vols. 10-11, Mansell, 1975-77; London: vol. 13, Taylor & Francis, 1986), 3:193, quoted in P. B. Wood, "Methodology and Apologetics: Thomas Sprat's *History of the Royal Society*," *The British Journal for the History of Science* 13 (1980):5.

4 John Dolben was dean of Westminster at the time of his election. His uncle was Gilbert Sheldon, Archbishop of Canterbury. Unless otherwise noted, the *Dictionary of National Biography* is the source of the biographical information in this chapter.

5 Samuel Pepys wrote in his diary for 10 February 1667, "Up and with my wife to church, where Mr. Mills made an unnecessary sermon upon Originall Sin, neither understood by himself nor the people." Samuel Pepys, *The Diary of Samuel Pepys*, ed. Robert Latham and William Matthews, 10 vols. (Berkeley: University of California Press, 1970-1983), 8:51. On 9 May 1669, Pepys noted that "Dr. Mills preached a dull sermon." Pepys, *Diary*, 9:548. In his entry for 19 January 1662, Pepys criticized another sermon he had heard Mills preach:

> "To church in the morning, where Mr. Mills preached upon Christ's being offered up for our sins. And there, proveing the aequity with what Justice God could lay our sins upon his Son, he did make such a sermon (among other things, pleading from God's universall Soverainty over all his Creatures, the power he hath of commanding what he would of his Son, by the same rule as that he might have made us all and the whole world from the beginning to have been in hell, arguing from the power of the potter hath over his clay), that I could have wished he had let it alone. And speaking again, that God the Father is now so satisfyd by our Security for our debt that we might say at the last day, as many of us as have interest in Christ's death—Lord, we owe thee nothing—our debt is paid—we are not beholden to thee for anything, for thy debt is paid thee to the full—which methinks were very bold words." Pepys, *Diary*, 3:12-13.

This was not the only sermon to receive his disapproval. On 9 August 1663, Pepys complained in his diary:

> "Up; and leaving my brother John to go somewhither else, I to church and heard Mr. Mills (who is lately returned out of the country, and it seems was fetched in by many of the parishioners with great state) preach upon the Authority of the Ministers, upon these words: 'Wee are therefore Embassadors of Christ.' Wherein, among other high expressions, he said that such a learned man used to say that if a minister of the word and an Angell should meet him together, he would salute the Minister first—which methought was a little too high." Pepys, *Diary*, 4:268.

See also Pepys, *Diary*, 6:131, for his remarks upon Mills's "sorry sermon" of 18 June 1665.

6 Biographical details from Webster, *Great Instauration*, pp. 83-84.

7 John Aubrey, *"Brief Lives," chiefly of contemporaries, set down by John Aubrey, between the years 1669 & 1696*, ed. Andrew Clark, 2 vols. (Oxford: Clarendon Press, 1898), 2:127. Biographical details are taken from Charles Webster, ed., *Samuel Hartlib and the Advancement of Learning* (Cambridge: Cambridge University Press, 1970), pp. 14-16.

8 See I. M. Green, *The Re-establishment of the Church of England: 1660-1663* (New York: Oxford University Press, 1978), p. 118.

9 Norman Sykes, *From Sheldon to Secker: Aspects of English Church History, 1660-1768*, The Ford Lectures [1958] (Cambridge: Cambridge University Press, 1959), p. 109.

10 Edward Waterhouse, *A Discourse and Defence of Arms and Armory, Shewing The Nature and Rises of Arms and Honour in England, from the Camp, the Court, the City: Under The two later of which, are contained Universities and Inns of Court* (London: T.R. for Samuel Mearne in Little Britain, 1660), p. 133.

11 Edward Waterhouse, *Fortescutus Illustratus, or a Commentary On that Nervous Treatise De Laudibus Legum Angliae, Written by Sir John Fortescue Knight, First Lord Chief Justice, after Lord Chancellour to King Henry the Sixth. Which Treatise, dedicated to Prince Edward that King's Son and Heir (Whom he attended in his retirement into France, and to Whom he loyally and affectionately imparted Himself in the Virtue and Variety of His Excellent Discourse) Hee purposely wrote to consolidate his Princely minde in the love and approbation of the good Lawes of England, and of the laudable Customs of this his Native Country* (London: Printed by Tho. Roycroft for Thomas Dicas at the Sign of the Hen and Chickens in St. Paul's Church-yard, 1663).

12 Edward Waterhous[e], *An humble Apologie for Learning and Learned Men* (London: Printed by T. M. for M. M. G. Bedell, and T. Collins, at the middle Temple in Fleetstreet, 1653). For a good account of the Interregnum debate over reform of the universities, see Allen G. Debus, *Science and Education in the Seventeenth Century: The Webster-Ward Debate* (New York: American Elsevier Inc., 1970), pp. 1-64.

13 Webster, *Great Instauration*, p. 450, quoting from one of four letters: Robert Wood to William Potter, 1 December 1658, 15 January 1658/[9], 2 April 1659, or 25 May 1659, Hartlib Papers, bundle no. xxxiii, section 1, Sheffield University Library, Sheffield, England.

14 Thomas Sprat reports that when the King was informed of the election of John Graunt, the King "gave this particular charge to His Society, that if they found any more such Tradesmen, they should be sure to admit them all, without more ado." John Graunt, the statistician, was "a Shop-keeper of *London*." See Thomas Sprat, *The History of the Royal-Society of London, For the Improving of Natural Knowledge* (London: Printed by T.R. for J. Martyn at the Bell without Temple-bar, and J. Allestry at the Rose and Crown in Duck-lane, Printers to the Royal Society, 1667; reprint ed., ed. Jackson I. Cope and Harold Whitmore Jones, St. Louis: Washington University Studies, 1958), p. 67.

15 Anthony à Wood, *Athenae Oxonienses. An Exact History of All the Writers and Bishops Who Have Had Their Education in the University of Oxford. To Which Are Added, The Fasti, or Annals of the Said University*, ed. Philip Bliss, 3rd. ed., 4 vols. (London: Printed for F. C. and J. Rivington; Lackington, Allen, and Co.; T. Payne; White, Cochrane, and Co.; Longman, Hurst, Rees, Orme, and Brown; Cadell and Davies; J. and A. Arch; J. Mawman; Black, Parry, and Co.; R. H. Evans; J. Booth, London; and J. Parker, Oxford, 1813-20; facs. reprint ed., Anglistica & Americana, no. 22, Hildesheim: George Olms, 1969), 4:286.

16 Tho[mas] Baker, *The Geometrical Key: or the Gate of Equations Unlock'd: A New discovery of the Construction of all Equations, howsoever affected, not exceeding the fourth Degree; viz. of, Linears, Quadratics, Cubics, Biquadratics; And the finding of all their Roots, as well false, as true; without the use of Mesolabe, Trisection of Angles; without Reduction, Depression, or any other previous preparation of Equations, by a Circle, and any (and that but one only) Parabole* (London: Printed by J. Playford, for R. Clavel, at the Peacock in S. Paul's Church-yard, 1684).

17 Thomas Birch, *The History of the Royal Society of London for Improving of Natural Knowledge From its First Rise. In which The most considerable of those Papers communicated to the Society, which have hitherto not been published, are inserted in their proper order, as a supplement to The Philosophical Transactions*, 4 vols. (London: n.p., 1756-57; reprint ed., with an introduction by A. Rupert Hall and additional bibliographical note by Marie Boas Hall, New York: Johnson Reprint Co., The Sources of Science, no. 44, 1968), 4:527.

18 See Jos[eph] Glanvill, *Plus Ultra: Or The Progress and Advancement of Knowledge Since the Days of Aristotle. In an Account of some of the most Remarkable Late Improvements Of Practical, Useful Learning: To Encourage Philosophical Endeavours. Occasioned By a Conference with one of the Notional Way* (London: Printed for James Collins at the Kings-Head in Westminster-Hall, 1668; facs. reprint ed., edited by Jackson I. Cope, Gainesville, FL: Scholars' Facsimiles & Reprints, 1958), pp. 24-27. Page numbers cited are from the 1958 facsimile reprint edition. In citing the "*Pious*, *Learned*, and *Excellent* Persons" of the clergy who were members of the Society, Glanvill added that "mention of those Celebrated Names, may serve to remove another grondless suspicion which some have entertained, *viz.* That the *Universities* are undermined by this new *Philosophick Society*: For whoever phancieth or suggests *that*, casts a *black Character* upon the *sagacity* and *faithfulness* of those *Reverend* Men, who all have been *Eminent Members* of one or other of those *Schools* of *Learning*; and most of them do still retain a Relation to those *ancient* and *venerable* Bodies." Ibid., pp. 26-27. For an account of the hostility to which the Society was subject from university circles, see Margery Purver, *The Royal Society: Concept and Creation* (London: Routledge and Kegan Paul, 1967), pp. 70-75. See also R. H. Syfret, "Some Early Reactions to the Royal Society," *Notes and Records of the Royal Society of London* 7 (April 1950):219-228.

19 Sprat explained that the philosophic liberty which the Society professed was not opposed to the "retaining of *antient Truths*." Sprat, *History*, pp. 105-06. Sprat praised the "Criticks" and "Philologists" whose labors had uncovered "the Monuments of the *Antients*," and he distanced himself from those "*Assertors of new Philosophy*" who rejected "*Root* and *Branch*, of whatever has the face of *Antiquity*." Sprat, *History*, pp. 328-29, 24-25. Glanvill praised such works of Aristotle as his "*Rhetorick*, *History of Animals*, and *Mechanicks*," and claimed that "*Peripatetick Studies*," so long as they were tempered by a healthy dose of skepticism were, not "*disallowed* by the men of the *Practical Method*." Glanvill, *Plus Ultra*, pp. xxvii, 127-28.

20 Sprat observed of the Society that it was "in vain for any man amongst them to strive to preferr himself before another; or to seek for any great glory from the subtilty of his Wit; seeing it was the inartificial process of the *Experiment*, and not the *Acuteness* of any Commentary upon it, which they have had in veneration." Sprat, *History*, p. 91. Sprat maintained that the "calmness, and unpassionate evenness of the true Philosophick Spirit" contrasted sharply with the vainglory of the dogmatists. The latter were "prone to under-value other mens labours, and to neglect the real advantage, that may be gotten by their assistance least they should seem to darken their own glory." Sprat, *History*, p. 33. Sprat suggested that had the latter ever experienced all the difficulties that attended the true search

for knowledge, instead of "onely hastily catching things in small *Systems*," they would be less imperious and more critical of their own convictions. Sprat, *History*, p. 69.

21 Sprat declared that there could "never be found, in the breast of any particular *Philosopher* as much wariness, and coldness of thinking, and rigorous examination" as could be found among a group joined together in the pursuit of truth. Such a group was the Royal Society. By engaging in corporate research, its members, in Sprat's opinion, "guarded themselves against themselves, lest the strength of their own thoughts should lead them into error; lest their good Fortune in one Discovery, should presently confine them onely to one way of trial; lest their failings should discourage, or their success abate their diligence." Sprat, *History*, pp. 102, 92.

22 Barbara J. Shapiro, *Probability and Certainty in Seventeenth-Century England: A Study of the Relationships Between Natural Science, Religion, History, Law, and Literature* (Princeton: Princeton University Press, 1983), p. 13.

23 Ibid., p. 120.

24 Sir George Wheler, *A Journey into Greece By George Wheler Esq; In company of Dr Spon of Lyons. In Six Books. Containing I. A Voyage from Venice to Constantinople. II. An Account of Constantinople and the Adjacent Places. III. A Voyage through Lesser Asia. IV. A Voyage from Zant through several Parts of Greece to Athens. V. An Account of Athens. VI. Several Journeys from Athens, into Attica, Corinth, Boetia, &c. With variety of Sculptures* (London: Printed for William Cademan, Robert Kettlewell, and Awnsham Churchill, at the Popes Head in the New Exchange, the Hand and Scepter in Fleetstreet, and the Black Swan near Amen-Corner, 1682), p. 425.

25 John Wallis, *A Treatise of Algebra, Both Historical and Practical. Shewing, The Original, Progress, and Advancement thereof, from time to time; and by what Steps it hath attained to the Heighth at which now it is. With some Additional Treatises, I. Of the Cono-Cuneus; being a Body representing in part a Conus, in part a Cuneus. II. Of Angular Sections; and other things relating thereunto, and to Trigonometry. III. Of the Angle of Contact; with other things appertaining to the Composition of Magnitudes, the Inceptives of Magnitudes, and the Compositions of Motions, with the results thereof. IV. Of Combinations, Alternations, and Aliquot Parts* (London: Printed by John Playford, for Richard Davis, Bookseller, in the University of Oxford, 1685), pp. 280-355, cited in Margaret E. Baron, *The Origins of the Infinitesimal Calculus* (Oxford: Pergamon Press, 1969), p. 213.

26 Michael Sean Mahoney, *The Mathematical Career of Pierre de Fermat (1601-1665)* (Princeton: Princeton University Press, 1973), p. 352.

27 Scriba states that Wallis was the first to use the term, interpolation, to describe his method of mathematics. See Christoph J. Scriba, "John Wallis," in *Dictionary of Scientific Biography*, ed. Charles Coulston Gillispie, 16 vols. (New York: Charles Scribner's Sons, 1970-80), 14:148. For a discussion of Wallis's use of interpolation, see Richard S. Westfall, *Never at Rest: A Biography of Isaac Newton* (Cambridge: Cambridge University Press, 1980), pp. 33-38.

28 This point is made by Scriba in "Wallis," p. 149.

29 For Newton's debt to Wallis, see Westfall, *Never at Rest*, pp. 114-115, 122-23.

30 Westfall, *Never at Rest*, pp. 102-3, 33.

31 Westfall, *Never at Rest*, p. 222.

32 Westfall, *Never at Rest*, p. 258.

33 Westfall, *Never at Rest*, p. 102.

34 Westfall, *Never at Rest*, p. 202. The edition of *Logarithmo-technia* which Barrow lent Newton was presumably: Nicholas Mercator, *Logarithmo-technia: Sive Methodus construendi Logarithmos Nova, accurata, & facilis; scripto Antehac Communicata, Anno Sc. 1667. Nonis Augusti: Cui nunc accedit Vera Quadratura Hyperbolae, & Inventio Summae Logarithmorum. . . . Huic Etiam Jungitur Michaelis Angeli Riccii Exercitatio Geometrica de Maximis & Minimis; hic ob Argumenti praestantiam & Exemplarium raritatem recusa* (London: Typis Guilielmi Godbid, & Impensis Mosis Pitt Bibliopolae, in vico vulgo vocato Little Britain, 1668).

35 J. M. Child, Introduction to *The Geometrical Lectures of Isaac Barrow*, trans. and ed. J. M. Child (Chicago: The Open Court Publishing Company, The Open Court Series of Classics of Science and Philosophy, no. 3, 1916), pp. 16-20, 28-29, cited by Baron, *Calculus*, p. 252.

36 This is the opinion of Margaret Baron. See Baron, *Calculus*, pp. 251-52. See also Westfall, *Never at Rest*, p. 131, n. 55.

37 Add MS 3968.41, f. 86v, Portsmouth Papers, Cambridge University Library, quoted in Westfall, *Never at Rest*, p. 131, n. 55.

38 Westfall, *Never at Rest*, p. 30.

39 See Mahoney, *Fermat*, p. 31.

40 That algebra remained for Barrow a useful logical tool, not a branch of mathematics is the judgment of D. T. Whiteside, "Isaac Barrow," in *Dictionary of Scientific Biography*, ed. Charles Coulston Gillispie, 16 vols. (New York: Charles Scribner's Sons, 1970-80), 1:474.

41 Mahoney, *Fermat*, p. 337. For a discussion of the exchange of letters between Fermat and Wallis, see Mahoney, *Fermat*, pp. 330-37. The mathematics of the Pell equation is discussed in Mahoney, *Fermat*, pp. 319-23, and in P. J. Wallis, "John Pell," *Dictionary of Scientific Biography*, ed. Charles Coulston Gillispie, 16 vols. (New York: Charles Scribner's Sons, 1970-80), 10:495.

42 John Pell, "An Idea of Mathematics written by Mr Joh. Pell to Samuel Hartlib," in John Dury, *The Reformed Librarie Keeper With a Supplement to the Reformed-School, As subordinate to Colleges in Universities. Whereunto is added I. An Idea of Mathematicks. II. The description of one of the chiefest Libraries which is in Germanie, erected and ordered by one of the most Learned Princes in Europe* (London: Printed by William Du-Gard, and are to bee sold by Rob. Littleberrie at the sign of the Unicorn in Little Britain, 1650), p. 38. Charles Webster cites the same passage in *Great Instauration*, p. 358, but his quotation is taken from p. 38 of the second edition of Dury's *The Reformed School* (1651).

43 Wallis, "Pell," p. 495. Charles Webster reports that the 1638 Latin translation of Pell's "Idea," which Haak sent Mersenne, cannot be found. The original manifesto, which Pell wrote in English, has only recently come to light. See Webster, *Hartlib*, pp. 15n, 139-40; and P. J. Wallis, "An Early Mathematical Manifesto—John Pell's *Idea of Mathematics*," *Durham Research Review* no. 18 (April, 1967):139-48.

44 Wallis, "Pell," p. 495. See John Pell, *Controversiae de vera Circuli mensura Anno [1644] exortae, inter Christianum Severini Longomontanum, Cimbrum, Superiorum Mathematum in Regia Danorum Academia Hauniensi, Professoren publicum et Ioannem Pellium, Coritano-Regnum, Anglum, Matheseos, in Illustri Amstelodamensium Gymnasio, Professorem publicum. Pars Prima* (Amsterdam: Apud Ioannem Blaeu, 1647). According to E. J. Dijksterhuis, Pell was replying to *Christiani Severini Longomontani, Cimbri, Rotundi in plano, seu Circuli, absolut mensura, duobus libellis comprehensa, quorum prior*

veram constitutionem Peripheriae Circuli Synthetice perficit, et mox hujus ad Diametrum rationem (cui apologia adversus Ludolphaeos est adjecta) Posterior Geodaesiam Rotundi plano analytice absolvit (Amsterdam: Apud Ioan Blaeu, 1644). See n. 5 and n. 10 in E. J. Dijksterhuis, "John Pell in Zijn Strijd Over De Rectificatie van den Cirkel," *Euclides* 8 (1932):294-95.

45 See Aubrey, *Brief Lives*, 2:125.

46 Wallis, "Pell," p. 495.

47 The following account of Ward's equant method is drawn from J. L. Russell, "Kepler's Laws of Planetary Motion: 1609-1666," *The British Journal for the History of Science* 2 (1964):1-24, and Curtis A. Wilson, "From Kepler's Laws, So-called, to Universal Gravitation: Empirical Factors," *Archive for History of Exact Sciences* 6 (1970):117-21. Ward's method first appeared in Ward's *In Ismaelis Bullialdi Astronomiae Philolaicae Fundamenta, Inquisitio Brevis* (Oxford: Excudebat Leon. Lichfield Academiae Typographus, 1653). He discussed it again three years later in his *Astronomiae Geometricae Liber Tertius, Astronomia Circularis Geometrice Proposita* (London: Typis Jacobi Flesher, Prostant apud Cornelium Bee, 1656).

48 Russell, "Laws" p. 2.

49 *Historia Coelestis Britannica*, 3 vols. (London: Typis H. Meere, 1725); *Atlas Coelestis* (London: n.p., 1729). This appreciation of Flamsteed's achievement is drawn from Victor E. Thoren, "John Flamsteed," in *Dictionary of Scientific Biography*, ed. Charles Coulston Gillispie, 16 vols. (New York: Charles Scribner's Sons, 1970-80), 5:23-24.

50 See Thoren, "Flamsteed," pp. 23-24.

51 Jeremiah Horrocks, *Jeremiae Horroccii, Liverpoliensis Angli, ex Platinatu Lancastriae Opera Posthuma; viz. Astronomia Kepleriana, defensa & promota. Excerpta ex Epistolis ad Crabtraeum suum. Observationum Coelestium Catalogus. Lunae Theoria nova. Accedunt Guilielmi Crabtraei, Mancestriensis, Observationes Coelestes. In calce adjiciuntur Johannis Flamstedii, Derbiensis, De Temporis Aequatione Diatriba. Numeri ad Lunae Theoriam Horroccianam*, ed. John Wallis (London: Typis Gulielmi Godbid, Impensis J. Martyn Regalis Societatis Typographi, ad insigne Campanae in Coemeterio D. Pauli, 1673).

52 Thoren, "Flamsteed," pp. 23-24. Flamsteed revised his lunar theory for Part 2 of his *Doctrine of the Sphere* in 1680. See [John Flamsteed], *The Doctrine of the Sphere, Grounded on the Motion of the Earth, And the Antient Pythagorean or Copernican System of the World. In Two Parts* (London: Printed by A. Godbid and J. Playford, 1680), in Sir Jonas Moore, *A New Systeme of the Mathematicks: Containing I. Arithmetick, as well Natural and Decimal, as in Species, or the Principles of Algebra. II. Practical Geometry together with the first Six Books of Euclid's Elements, as also the Eleventh and Twelfth, symbolically demonstrated. III. Trigonometry Plain and Spherical. IV. Cosmography, or a Description of the Heavens. V. Navigation, or Sailing by a Plain or Mercator's Chart; as also by the Arch of a Great Circle, &c. VI. The Doctrine of the Sphere, grounded on the Motion of the Earth, according to the Old Pythagorean and Copernican Systeme. VII. Astronomical Tables, with Tables of Logarithms, Natural and Artificial Sines and Tangents, and Versed Sines. VIII. A New Geography, or a Description of the most Eminent Countries and Coasts of the World, with Maps of them, and Tables of their Latitude and Longitude* (London: Printed by A. Godbid and J. Playford for Robert Scott, Bookseller in Little Britain, 1681). Flamsteed revised his theory again in the latter 1690's and again around 1703. See Thoren, "Flamsteed," p. 24.

53 The following discussion of Wilkins's career and writings is heavily indebted to Hans Aarsleff, "John Wilkins," *Dictionary of Scientific Biography*, ed. Charles Coulston Gillispie, 16 vols. (New York: Charles Scribner's Sons, 1970-80), 14:361-81.

54 John Wilkins, *An Essay Towards a Real Character And a Philosophical Language* (London: For Sa: Gellibrand, and for John Martyn Printer to the Royal Society, 1668).

55 [John Wilkins], *The Discovery of a World in the Moone. Or, A Discourse Tending, To Prove, that 'tis probable there may be another habitable World in that Planet* (London: Printed by E. G. for Michael Sparke and Edward Forrest, 1638).

56 Aarsleff, "Wilkins," pp. 364-65.

57 The *Discourse* appeared together with a third printing of the *Discovery* in [John Wilkins], *A Discourse concerning A New world & Another Planet In 2 Bookes* (London: Printed for Iohn Maynard, & are to be sold at the George, in Fleetstreet near St. Dunstans Church, 1640). The inside title page reads: [John Wilkins], *A Discourse concerning A New Planet. Tending to prove, That 'tis probable our Earth is one of the Planets. The second Booke, now first published* (London: Printed by R. H. for Iohn Maynard, and are to be sold at the George in Fleetstreet, neer S. Dunstans Church, 1640).

58 John Wilkins, *Mercury, or the Secret and Swift Messenger: Showing, How a Man may with Privacy and Speed communicate his Thoughts to a Friend at any distance* (London: Printed by I. Norton, for Iohn Maynard, and Timothy Wilkins, and are to be sold at the George in Fleetstreet, neere Saint Dunstans Church, 1641).

59 Aarsleff, "Wilkins," p. 365.

60 I[ohn] W[ilkins], *Mathematical Magick. Or The Wonders That may be performed by Mechanicall Geometry. In two Books. Concerning Mechanicall {Powers. {Motions. Being One of The most easie, pleasant, usefull, (and yet most neglected) part of Mathematicks. Not before treated of in this language* (London: Printed by M. F. for Sa: Gellibrand at the brasen Serpent in Pauls Church-yard, 1648).

61 Ibid., p. 92, quoted in Aarsleff, "Wilkins," p. 366.

62 Aarsleff, "Wilkins," p. 366.

63 Webster, *Great Instauration*, p. 159.

64 Aarsleff, "Wilkins," p. 369.

65 Sprat, *History*, p. 94, quoted in Wood, "Methodology and Apologetics," p. 3.

66 John Ray, *Methodus Plantarum Nova, Brevitatis & Perspicuitatis Causa Synoptice in Tabulis Exhibita; Cum notis Generum tum summorum tum subalternorum Characteristicis, Observationibus nonnullis de seminibus Plantarum & Indice Copioso* (London: Impensis Henrici Faithorne, & Joannis Kersey ad insigne Rosae in Coemeterio D. Pauli, 1682).

67 Alfred Rupert Hall, *The Scientific Revolution, 1600-1800* (London: Longmans, Green and Co., 1954), p. 283.

68 [John Ray], *Catalogus Plantarum Circa Cantabrigiam nascentium: In quo exhibentur Quotquot hactenus inventae sunt, quae vel sponte proveniunt, vel in agris feruntur; Una cum Synonymis selectioribus, locis natalibus & observationibus quibusdam oppido raris. Adjiciuntur in gratiam tyronum, Index Anglico-latinus, Index locorum, Etymologia nominum, & Explicatio quorundam terminorum* (Cambridge: Excudebat Joann. Field, celeberrimae Academiae Typographus. Impensis Gulielmi Nealand, Bibliopolae, 1660).

69 Ellison Hawks, *Pioneers of Plant Study* (n.p., 1928; reprint ed., Freeport, NY: Books for Libraries Press, Essay Index Reprint Series, 1969), p. 62.

70 Ray's *Historia Plantarum* was published in three volumes. Volume 1: John Ray, *Historia Plantarum Species hactenus editas aliasque insuper multas noviter inventas & descriptas complectens. In qua agitur primo De Plantis in genere, Earumque Partibus, Accidentibus & Differentiis; Deinde Genera omnia tum summa tum subalterna ad Species usque infimas, Notis suis certis & Characteristicis Definita, Methodo Naturae vestigiis insistente disponuntur; Species singulae accurate describuntur, obscura illustrantur, omissa supplentur, superflua resecantur, Synonyma necessaria adjiciuntur; Vires denique & Usus recepti compendio traduntur* (London: Typis Mariae Clark: Prostant apud Henricum Faithorne Regiae Societatis Typographum ad Insigne Rosae in Coemeterio D. Pauli, 1686). Volume 2: John Ray, *Historiae Plantarum Tomus Secundus: Cum Duplici Indice Generali Altero Nominum & Synonymorum praecipuorum; Altero Affectuum & Remediorum: Accessit Nomenclator Botanicus Anglo-Latinus* (London: Typis Mariae Clark: Prostant apud Henricum Faithorne Regiae Societatis Typographum, ad insigne Rosae in Coemeterio D. Pauli, 1688). Volume 3: John Ray, *Historiae Plantarum Tomus Tertius: qui est Supplementum Duorum praecedentium: Species omnes vel omissas, vel post Volumina illa evulgata editas, praeter inumeras fere novas & indictas ab Amicis communicatas complectens: cum Synonymis necessariis, et Usibus in Cibo, Medicina, & Mechanicis: Additio ad Opus consummandum Generum Indice copioso. Accessit Historia Stirpium Ins. Luzonis & reliquarum Philippinarum a R.P. Geo. Jos. Camello, Moravo-Brunensi, S.J. conscripta. Item D. Jos. Pitton Tournefort, M.D. Parisiensis, & in Horto Reg. Botanices Professoris, Corollarium Institutionum Rei Herbariae* (London: Apud Sam. Smith & Benj. Walford, Reg. Soc. Typographos, ad Insignia Principis in Area Boreali D. Pauli, 1704).

71 Francis Willughby, *Francisci Willughbeii De Middleton in agro Warwicensi, Armigeri, E Regia Societate, Ornithologiae Libri Tres: In quibus Aves omnes hactenus cognitae in methodum naturis suis convenientem redactae accurate describuntur, Descriptiones Iconibus elegantissimis & vivarum Avium simillimis, AEri incisis illustrantur*, ed. John Ray (London: Impensis Joannis Martyn, Regiae Societatis Typographi, ad insigne Campanae in Caemeterio D. Pauli, 1676) and Francis Willughby, *De Historia Piscium Libri Quatuor, Jussu & Sumptibus Societatis Regiae Londinensis editi. In quibus non tantum De Piscibus in genere agitur, Sed & species omnes, tum ab aliis traditae, tum novae & nondum editae bene multae, naturae ductum servante Methodo dispositae, accurate describuntur. Earumque effigies, quotquot haberi potuere, vel ad vivum delineatae, vel ad optima exemplaria impressae; Artifici manu elegantissime in AEs incisae, ad descriptiones illustrandas exhibentur. Cum Appendice Historias & Observationes supplementum Operis collatas complectente*, ed. John Ray (Oxford: E Theatro Sheldoniano, 1686).

72 J[ohn] Ray, *A Collection of English Proverbs Digested into a convenient Method for the speedy finding any one upon occasion; with Short Annotations. Whereunto are added Local Proverbs with their Explications, Old Proverbial Rhythmes, Less known or Exotick Proverbial Sentences, and Scottish Proverbs* (Cambridge: Printed for John Hayes, Printer to the University, for W. Morden, 1670).

73 John Ray, *A Collection of English Words, Not Generally used, with their Significations and Original, in two Alphabetical Catalogues, The one of such as are proper to the Northern, the other to the Southern Counties. With Catalogues of English Birds and Fishes: And an Account of the preparing and refining such Metals and Minerals as are gotten in England* (London: Printed by H. Bruges. For Tho. Burrell at the Golden-Ball under St. Dunstans Church in Fleetstrete, 1674).

74 John Ray to Edward Lhwyd, 25 November 1691, in *Further Correspondence of John Ray*, ed. Robert W. T. Gunther (London: Printed for the Ray Society, 1928), p. 222, quoted in

Charles E. Raven, *John Ray, Naturalist: His Life and Works* (Cambridge: Cambridge University Press, 1942), p. 430. Ray made this claim in the course of telling Lhwyd that he had finished the *Miscellaneous Discourses* and sent it to the printer. See John Ray, *Miscellaneous Discourses Concerning the Dissolution and Changes of the World Wherein The Primitive Chaos and Creation, the General Deluge, Fountains, Formed Stones, Sea-Shells found in the Earth, Subterraneous Trees, Mountains, Earthquakes, Vulcanoes, the Universal Conflagration and Future State, are largely Discussed and Examined* (London: Printed for Samuel Smith, at the Prince's Arms in St. Paul's Church Yard, 1692). For the second edition of this work, see: John Ray, *Three Physico-Theological Discourses, Concerning I. The Primitive Chaos, and Creation of the World. II. The General Deluge, its Causes and Effects. III. The Dissolution of the World, and Future Conflagration. Wherein Are largely Discussed the Production and Use of Mountains; the Original of Fountains, of Formed Stones, and Sea-Fishes, Bones and Shells found in the Earth; the Effects of particular Floods and Inundations of the Sea; the Eruptions of Vulcano's; the Nature and Causes of Earthquakes: With an Historical Account of those Two late Remarkable Ones in Jamaica and England. With Practical Inferences*, 2nd. ed. (London: Printed for Sam. Smith, at the Princes Arms in St. Paul's Church-yard, 1693).

75 John Ray, *The Wisdom of God Manifested in the Works of the Creation. Being the Substance of some common Places delivered in the Chappel of Trinity-College, in Cambridge* (London: Printed for Samuel Smith, at the Princes Arms in S. Pauls Church-Yard, 1691), preface.

76 Raven, *Ray*, p. 452.

77 K. Theodore Hoppen, *The Common Scientist of the Seventeenth Century: A Study of the Dublin Philosophical Society, 1683-1708* (Charlottesville: University Press of Virginia, 1970), pp. 36-37.

78 St. George Ashe, "Concerning the Squaring of the Circle," Add. MS 4811, ff. 53-7, British Museum, London, cited by Hoppen, *Common Scientist*, pp. 118-19.

79 Hoppen, *Common Scientist*, p. 103.

80 Hoppen, *Common Scientist*, p. 64.

81 Hoppen, *Common Scientist*, p. 114. Clouds spoiled the results of the Dublin group. See "Observations of the Solar Eclipse July the 2. 1684. at Oxford, sent in a Letter from Dr. Edw. Bernard, Astron. Prof. Oxon. to Mr. John Flamsteed, Astron. Reg. At Lisbon by Mr. Jacobs; at Dublin, by Mr. Ash and Mr. Molyneaux; and at Tredagh by Mr. Osborne," *Philosophical Transactions* 14 (20 October 1684):749.

82 See Hoppen, *Common Scientist*, pp. 131-34.

83 St. George Ashe, "A Discourse of the Air," Add. MS 4811, ff. 135v-8, British Museum, London, and The Minutes of the Dublin Philosophical Society, 25 October 1686, Add. MS 4811, f. 180v, British Museum, London, quoted in Hoppen, *Common Scientist*, p. 133.

84 [William Oughtred], *Arithmeticae In Numeris Et Speciebus Institutio: Quae Tum Logisticae, Tum Analyticae, Atque Adeo Totius Mathematicae, Quasi Clavis Est* (London: Apud Thomam Harperum, 1631).

85 Webster, *Great Instauration*, p. 416. Wood's help is acknowledged in the preface (viz., "To the Reader") of Will[iam] Oughtred, *The Key of the Mathematicks New Forged and Filed: Together with A Treatise of the Resolution of all kinde of Affected Aequations in Numbers. With the Rule of Compound Usury; And demonstration of the Rule of false Position. And a most easie Art of delineating all manner of Plaine Sun-Dyalls* (London: Printed by Tho. Harper, for Rich. Whitaker, and are to be sold at his shop in Pauls Church-

yard, 1647). The edition which Ward and Wallis helped to prepare is William Oughtred, *Clavis Mathematicae Denuo Limata, Sive potius Fabricata. Cum aliis quibusdam ejusdem Commentationibus, quae in sequenti pagina recensentur*, 3rd. ed. (Oxford: Excudebat Leon. Lichfield, Veneunt apud Tho. Robinson, 1652).

86 R[obert] W[ood], "TEN TO ONE or A short and ready way For the extraordinary Facilitation & Dispatch of RECKONINGS. By meanes only of Two or Three NEW DECIMAL COINES much desired in order to The Publick Good," in Webster, *Great Instauration*, pp. 535-39.

87 Webster, *Great Instauration*, pp. 416-20.

88 Webster, *Great Instauration*, p. 452. See also pp. 450-52.

89 [William Petty], *A Treatise of Taxes & Contributions. Shewing the Nature and Measures of {Crown-Lands. {Assessements. {Customs. {Poll-Moneys. {Lotteries. {Benevolence. {Penalties. {Monopolies. {Offices. {Tythes. {Raising of Coins. {Harth-Money. {Excize, &c. With several intersperst Discourses and Digressions concerning {Warres. {The Church. {Universities. {Rents & Purchases. {Usury & Exchange. {Banks & Lombards. {Registries for Conveyances. {Beggars. {Ensurance. {Exportation of {Money. {Wool} {Free-Ports. {Coins. {Housing. {Liberty of Conscience, &c. The same being frequently applied to the present State and Affairs of Ireland* (London: Printed for N. Brooke, at the Angel in Cornhill, 1662), pp. 30-33, cited by Webster, *Great Instauration*, p. 453.

90 Webster, *Great Instauration*, p. 433.

91 J[ohn] B[eale], *Herefordshire Orchards, A Patterne For all England. Written in an Epistolary Address to Samuel Hartlib Esq.* (London: Printed by Roger Daniel, 1657), p. 9, quoted in Webster, *Great Instauration*, p. 479.

92 John Beale to Henry Oldenburg, 31 March 1675, quoted in Birch, *History*, 3:217.

93 Robert Gregg Frank, Jr., "Science, Medicine and the Universities of Early Modern England: Background and Sources," *History of Science* 11 (1973):199.

94 Ibid., p. 202.

95 Ibid., p. 199.

96 See "Table 3: Oxford Scientists and Virtuosi, 1640-1675," in Robert Gregg Frank, Jr., *Harvey and the Oxford Physiologists: A Study of Scientific Ideas and Social Interaction* (Berkeley: University of California Press, 1980), pp. 63-89.

97 Frank, "Science," pp. 257-61

98 Bernard to Oldenburg, 14 September 1672, in Oldenburg, *Correspondence*, 9:242.

99 Thomas Smith, who shared Bernard's interest in the recovery of ancient knowledge, published Bernard's catalogue of extant manuscript sources for the writings of the ancient mathematicians. See Thomas Smith, *Admodum Reverendi & Doctissimi Viri, D. Roberti Huntingtoni, Epistolae: et Veterum Mathematicorum, Graecorum, Latinorum, & Arabum, Synopsis: Collectore Viro Clarissimo & Doctissimo, D. Edwardo Bernardo, Astronomiae in Academia Oxoniensi Professore Saviliano. Praemittuntur D. Huntingtoni & D. Bernardi Vitae* (London: Typis Gul. Bowyer, Impensis A. & J. Churchill, ad Insigne Nigri Cygni in Vico vulgo vocato Paternoster-Row, 1704).

100 Walter Pope, *The Life of the Right Reverend Father in God Seth, Lord Bishop of Salisbury, and Chancellor of the Most Noble Order of the Garter. With a brief Account of Bishop Wilkins, Mr. Lawrence Rooke, Dr. Isaac Barrow, Dr. Tubervile, And others* (London:

Printed for William Keblewhite, at the Swan in St. Paul's Church-yard, 1697; reprint ed., ed. J. B. Bamborough, Oxford: Basil Blackwell for the Luttrell Society, 1961), p. 91; Birch, *History*, 2:340.

101 Michael C. W. Hunter, *John Aubrey and the Realm of Learning* (London: Gerald Duckworth & Co. Ltd., 1975), p. 43.

102 Pigot's name appears along with those of Bathurst, Bernard, Smith, and Wallis in the list drawn up by Robert Plot of the founding members of the Oxford Philosophical Society. See Wood, *Athenae Oxonienses*, 1:xcviii.

103 See Tho[mas] Pigot, "An Account of the Earthquake that happened at Oxford and the parts adjacent Sept. 17. 1683. by a fellow of a College in that University, and of the Royal Society," *Philosophical Transactions* 13 (20 September 1683):311-21.

104 Tho[mas] Smith, *A Sermon of the Credibility of the Mysteries of the Christian Religion. Preached Before a Learned Audience* (London: Printed by Tho. Roycroft for Ric. Davis Bookseller in Oxford, 1675).

105 Wallis to Oldenburg, February 1, 1667/8, in Oldenburg, *Correspondence*, 4:141.

106 Thomas Smith, "A conjecture about an under-Current at the Streights-Mouth read before the Oxford Society, Dec. 21. 1683," *Philosophical Transactions* 14 (April 20, 1684):564-66.

– 3 –

THE ACTIVITY OF THE CLERICAL VIRTUOSI IN THE ROYAL SOCIETY

The Royal Society both contributed to and was a social manifestation of that effervescence of interest in science that overtook England during the latter half of the seventeenth century. Through the cooperative efforts of its members the Royal Society sought to achieve an improvement and expansion in human knowledge concerning nature. The Society soon discovered that its mission stood in jeopardy if its members were not willing to contribute financially. Their financial cooperation became as important to the Society as their ability to contribute observations and experiments. Although Charles II declared himself the Society's founder and patron by its second charter in 1663, the Society did not receive significant income from the crown. Its straitened circumstances forced the Society to expand its rolls well beyond the fifty-five originally envisaged by its founders to include many who had only a superficial interest in the "new philosophy." In its search for patrons, the Royal Society did not overlook the Church of England.

The Election of Clergy to the Royal Society

The Royal Society elected four hundred twenty-six fellows between April 1663, and December 1687.[1] Forty-eight were clerics at the time of their election.[2] This includes the eight who were "original fellows" and the forty who joined after the signing of the second charter on May 20th, 1663. Together they comprised 9.4 percent of the total number of fellows elected in this period. This is a sizable percentage. It compares favorably with the percentages attained by other groups within the Society. Michael Hunter reports that courtiers and politicians were chosen in 15 percent of all elections until the end of 1685, aristocrats in 13 percent, gentlemen in 14 percent, merchants in 6 percent, and lawyers in 4 percent.[3]

It is important to remember that at no time did actual enrollments ever approach a number like four hundred twenty-six. The membership lists which the Society compiled annually for each anniversary-day meeting contained considerably fewer fellows. The number they averaged was one hundred ninety-six though some years saw the total descend as low as one hundred thirty-one fellows or rise as high as two hundred twenty-three.[4] On any given date, clerics composed anywhere from seven to fourteen percent of the total enrollment.

Tables 3.1 and 3.2 give a rough estimate of the fluctuations in membership from the signing of the second charter on the 20th of May 1663, to the end of December 1687.

TABLE 3.1

ELECTION, ATTRITION, AND CLERICAL ENROLLMENTS

Date	Total Enrolled	+	Number Elected	−	Attrition	=	Total Enrolled	Date
5/1663	8		3		0		11	11/1663
11/1663	11		9		0		20	12/1665
12/1665	20		10 + 2*		0 + 1*		31	11/1671
11/1671	31		5 + 1*		5		32	12/1675
12/1675	32		5		4		33	1/1681
1/1681	33		6 + 2*		13 + 2*		26	11/1685
11/1685	26		2		5		23	12/1687

SOURCES: Catalogue of Fellows 1660-1685 in Michael Cyril William Hunter, "The Social Basis and Changing Fortunes of an Early Scientific Institution," *Notes and Records of the Royal Society of London* 31 (July 1976):97-114; Chronological Register of Fellows in Royal Society of London, *The Record of The Royal Society of London*, 3rd. ed. (London: Oxford University Press, 1912), pp. 375-516; Officers and Fellows of the Royal Society 1663-1687 in Thomas Birch, *The History of the Royal Society*, 4 vols. (London: n.p., 1756-57; reprint ed., New York: Johnson Reprint Co., The Sources of Science, no. 44, 1968) 1:xxxiii-xlv; and *Dictionary of National Biography*.

*Not ordained at time of election.

TABLE 3.2

ELECTION, ATTRITION, AND ENROLLMENTS FOR THE SOCIETY AS A WHOLE

Date	Total Enrolled	+	Number Elected	-	Attrition	=	Total Enrolled	Date
5/1663	115		22		1		136	11/1663
11/1663	136		51		4		183	12/1665
12/1665	183		87		27		243	11/1671
11/1671	243		28		45		226	12/1675
12/1675	226		53		33		246	1/1681
1/1681	246		58		99		205	11/1685
11/1685	205		13		16		202	12/1687

SOURCES: Catalogue of Fellows 1660-1685 in Michael Cyril William Hunter, "The Social Basis and Changing Fortunes of an Early Scientific Institution," *Notes and Records of the Royal Society of London* 31 (July 1976):97-114; and Chronological Register of Fellows in Royal Society of London, *The Record of The Royal Society of London*, 3rd. ed. (London: Oxford University Press, 1912), pp. 375-516.

NOTE: The figures under "Total Enrolled" in Table 3.2 are approximate figures. They are calculated by adding the number of fellows elected within the time frames given and subtracting from those totals the number of fellows who died, resigned, or were expelled during the same time frames. The figures under "Number Elected" are derived from the *Record*; the figures under "Attrition" from the expulsion/resignation column of Hunter's catalogue.

Two variables determined both the number of clerics who were currently enrolled and their percentage of the total enrollment. The first was the rate at which they were elected to the Society. The second was the rate at which deaths, resignations, and expulsions removed them from the rolls. More than half of the clerical fellows elected between 1663 and 1685, entered the Society before 1671. Of these, thirteen were dead, three had been expelled, and one had resigned by the time the Society took a tally of its members in 1685. Of the sixteen clerics elected after 1671, the record is somewhat better. Only two had been dropped from the rolls by 1685. This post-1671 infusion of clerics was enough to offset the losses due to deaths, expulsions, and resignations until the beginning of 1681.

Between 1671 and 1681, the number of clerical fellows currently enrolled hovered between thirty and thirty-three.

This period of stable enrollments stood in sharp contrast to the situation before 1671. In the 1660's, clerics had joined the Society at a rate that was proportionally greater than the rate at which the Society itself was growing. As a result, the percentage of clerical members in the Royal Society had jumped from 7 percent in May of 1663, to 12.8 percent in November of 1671. This percentage continued to climb even after 1671, though it owed its rise to the disparity that developed between the rate of attrition which characterized clerical virtuosi as a group and that which characterized the Society as a whole.

Until the end of 1675, the clerical virtuosi were more resistant to the erosion wrought by death, expulsion, and resignation than the Society at large. After this date the tables were turned, and their proportion of the Society began to decline. Though the rate of clerical elections remained sufficient to maintain their numbers against attrition until 1681, it could not match the percentage by which fellows in general were being elected to the Society.

After 1681, attrition overwhelmed the ability of the clerical virtuosi to replenish their numbers through elections. The number of clerics enrolled plunged from thirty-three in 1681 to twenty-six in 1685. The severity of this decline should not surprise us even though it was proportionally higher than that suffered by the Society as a whole. The ranks of the clerical virtuosi contained a considerable number of older clerics. More than a third of the clerical virtuosi were forty-three or older when the second charter was signed in 1663. Expulsions also took their toll. Three clerics were stricken from the rolls in 1682, and four in 1685, for not paying their subscriptions.

It seems fair to conclude that the rate of election was the dominant factor determining enrollments before 1675, whereas the rate of attrition was the dominant factor after 1675. Enrollments stabilized in the 1670's because the rate at which clerics were being elected to the Society slowed drastically after the 1660's. In the 1660's, proportionally more fellows were elected in this subgroup than were elected in the Society as a whole. This continued to remain true until the end of 1675. Declines in the rate of election were not the only reason why enrollments in the early 1670's did not rise further than they did. Attrition first became a significant factor in this period. After 1675, it removed proportionally more members of the clerical subgroup than it did from the Society at large. The clerical election rate in the 1670's was able to replace the numbers that attrition removed, but its ability to do so was completely overwhelmed when the rate of attrition jumped in the early 1680's.

By and large, fluctuations in clerical enrollments appear to have followed fluctuations in enrollments for the Society as a whole. They mirror, for example, the rapid increases in enrollments that the Society experienced in the 1660's. For both, the highest percentage by which enrollments increased occurred in the same period, from November 1663, to December 1665. The same is true for the greatest decline in enrollments. This occurred in the period between January 1681, and November 1685. In the 1670's the fluctuations in their respective enrollments were more contrary. In the early 1670's, clerical enrollments did not go down when those of the Society at large went down. The percentage by which clerical enrollments increased did decline, however, from the previous period. When the Society's enrollments recovered in the late 1670's, those of the clerical virtuosi never rose by as high a percentage as those of the Society as a whole.

It is tempting to conjecture that Restoration politics had something to do with these fluctuations in enrollments. If they did, the clerical virtuosi were no more sensitive to their influence than the Society at large. When the election rate for clerical virtuosi is compared with the general election rate for the Society as a whole, the two do appear to rise and fall in direct proportion to the degree that the monarchy could command a majority in Parliament—an ability that was directly dependent upon the degree of political excitement in the nation. Yet it is difficult to measure the significance of this correlation since it could be coincidental.

The Extent of Clerical Participation in the Royal Society

At this juncture the nature of the sources precludes an exact determination of the attendance of clerical members at meetings of the Royal Society. Though Thomas Birch's *History of the Royal Society* gives figures for attendance at the meetings of the Council (i.e., the Society's administrative body), attendance at the general sessions of the Society remains a matter of informed conjecture.[5] Using G. E. Scala's index, Birch's *History* has been combed for instances where the activity of a cleric betrayed his presence at a meeting.[6] The result of compiling attendance figures in this way is to weight the statistics heavily in favor of those who made contributions, however small, to the intellectual life of the Society.

This results in counting the undemonstrative among the absent. Esse est percipi (apud Journalam).[7]

How grave a distortion does this procedure produce? It seems likely that the gap between actual attendance and the attendance figures that can be derived from Birch's *History* (viz. the attendance of those who contributed to the life of the Society) narrows after the first decade of the Society's existence. Sir Henry Lyons and and Dorothy Stimson report that attendance after 1670, dropped precipitously, rising only a bit during the early 1680's under the presidency of Wren. They give no figures, but they do cite Evelyn's letter to Pepys in June 1680, begging him for "one half-hour of your presence and assistance toward the most material concern of a Society which ought not to be dissolved for want of redress..."[8] Birch drops an occasional numerical total in his *History*, but more often than not the totals are of the anniversary-day meetings which normally drew larger crowds than regular meetings. It seems reasonable to expect that even if exact attendance figures were available, the proportion between those who regularly attended and those who did not would remain substantially the same—only the proportion of those who attended no meeting after their admission to the Society would change drastically. Michael Hunter has made the suggestion that we look to see whether fellows judged inactive by the criterion of Birch's *History* paid their dues regularly.[9] If they did, it suggests their attendance at the Society's meetings. Roger Williams, Gilbert Burnet, and John Tillotson paid their dues regularly for some period of time before falling into arrears, though Birch's *History* records no instance of their being in attendance at any meeting of the Royal Society. The same holds true for Edmund Castell who paid his dues regularly throughout his lifetime.[10]

Any informed conjecture has a margin of error. It has not always been easy to determine whether a reference in Birch's *History* implied that the cleric named was in attendance. Too often the proceedings of the Society are couched in the passive tense. That someone "was desired to do" something, does not indicate whether that someone was actually present at the time the order was given. Examples from the council minutes show that it was quite possible to be appointed to a committee without being present to decline or accept the honor.[11] This holds true for elections too. Even at the anniversary-day meetings, it is dangerous to assume that a cleric was present because he was elected to an office, unless the minutes record he was sworn in. There are many cases where a paper

TABLE 3.3

CLERGY WHO NEVER ATTENDED A MEETING AFTER THEIR ADMISSION TO THE ROYAL SOCIETY

Name	Date of Election
St. George Ashe	2/1686
Thomas Baker*	11/1684
John Beale	8/1663
Edward Bernard	4/1673
Gilbert Burnet	3/1664
Edmund Castell	4/1674
Henry Clerke	10/1667
John Dolben**	3/1665
Joseph Glanvill	12/1664
Humphrey Henchman	3/1665
Charles Hotham	1/1668
Henry Jenkes	11/1674
Benjamin Laney*	11/1666
Henry More	5/1664
George Morley	11/1666
Samuel Parker	6/1666
Francis Potter	11/1663
John Rogers*	4/1681
Gilbert Sheldon	3/1665
Richard Stearne	3/1665
John Tillotson	1/1672
Roger Williams	2/1664
Samuel Woodford	11/1664
Benjamin Woodroffe	4/1668
William Wotton**	12/1687

SOURCES: Thomas Birch, *The History of the Royal Society,* 4 vols. (London: n.p., 1756-57; reprint ed., New York: Johnson Reprint Co., The Sources of Science, no. 44, 1968); Gail Ewald Scala, "An Index of Proper Names in Thomas Birch, *The History of the Royal Society* (London, 1756-57)," *Notes and Records of the Royal Society of London* 28 (April 1974):263-329; Royal Society of London, *The Signatures in the First Journal-Book and the Charter Book of the Royal Society. Being a facsimile of the signatures of the founders, patrons and fellows of the Society from the year 1660 down to the present time* (London: Oxford University Press, 1936); and Catalogue of Fellows 1660-1685 in Michael Cyril William Hunter, "The Social Basis and Changing

Fortunes of an Early Scientific Institution," *Notes and Records of the Royal Society of London* 31 (July 1976):97-114.

*Though elected, Baker, Laney, and Rogers were never admitted. Fellows were usually admitted the first meeting they attended after their election. To be admitted, a fellow was presented to the Society by the president and invited to sign the Charter-book.

**The index prepared for the facsimile edition of the Charter-book lists Dolben as never having signed, but his name clearly appears on page 6. Though an active member, Wotton must be listed among the inactive since he does not begin to attend meetings until after December 1687, our cut-off date.

TABLE 3.4

CLERGY IN TABLE 3.3 WHO CORRESPONDED WITH THE ROYAL SOCIETY

Name	Number of Times Birch's *History* Records a Cleric Corresponding with the Royal Society
Ashe	7
Baker	2
Beale	19
Bernard	2
Glanvill	5
Potter	2
Rogers	1
Sheldon	1
Stearne	1
Tillotson	2

SOURCES: Thomas Birch, *The History of the Royal Society,* 4 vols. (London: n.p., 1756-57; reprint ed., New York: Johnson Reprint Co., The Sources of Science, no. 44, 1968), and Gail Ewald Scala, "An Index of Proper Names in Thomas Birch, *The History of the Royal Society* (London, 1756-57)," *Notes and Records of the Royal Society of London* 28 (April 1974):263-329.

was read though the author was absent—enough cases to make one wary of the passive construction. In compiling these attendance figures doubtful cases have been counted in favor of attendance. G. E. Scala's index of proper names in Birch's *History* also presents problems. Her index notes names, not titles. She does not completely exclude titles from her list; Sheldon often gets the credit where only the archbishop of Canterbury appears in the text. Yet it is not safe to assume this.

On the basis of evidence in Birch's *History*, Table 3.3 lists the clerical fellows who appear to have never attended a meeting after their admission to the Society. Of the twenty-four listed in Table 3.3, ten corresponded with the Royal Society. These ten are listed in Table 3.4.

Excluding the eight original clerical fellows who appear in the second charter, there remain another twenty who attended at least one meeting of the Society. Their names are given in Table 3.5. Three quarters of them also corresponded with the Society. The names of the eight who served on the Council are given in Table 3.6. Tables 3.3, 3.4, 3.5, and 3.6 suggest six ways of categorizing the activity of the 45 clerics who joined the Society after the signing of the second charter. These six ways form the headings of Table 3.7. The twenty-five clerics who never attended a meeting after their admission are divided into three groups: those who were completely inactive, those who occasionally corresponded, and those who corresponded regularly. For the purposes of this study 'attendance' is considered a higher order of activity than 'correspondence'. That is why the remaining twenty clerics are classified in terms of their attendance and not their correspondence. These clerics fall into two groups. Some attended sporadically; others attended regularly. None of the latter, excluding those listed under 'long term interest,' were able to sustain their level of attendance for longer than two and a half years. The last category takes in those clerics whose attendance or correspondence was so sustained as to indicate a deep and abiding interest in the Society. This category is evaluative in the sense that it measures something more subjective than just 'attendance' or 'correspondence.'

It is important to remember that Table 3.7 measures activity, not scientific accomplishment. The two were not necessarily equivalent. Smith, Vossius, and Pigot published in the *Philosophical Transactions*; Payne, Wheler, and Harrington did not. Yet the latter were more active in the proceedings of the Society than

TABLE 3.5

CLERGY ELECTED TO THE ROYAL SOCIETY AFTER THE SIGNING OF THE SECOND CHARTER WHO ATTENDED AT LEAST ONE MEETING AFTER THE DATE OF THEIR ADMISSION

Name	Date of Election	Number of Times Birch's *History* Records the Attendance of a Cleric
James Arderne* **	6/1668	1
Ralph Bathurst	8/1663	1
John Flamsteed**	2/1677	15
Thomas Gale**	12/1667	74
William Harrington	3/1666	1
Anthony Horneck*	1/1669	2
John Mapletoft**	2/1676	8
Daniel Milles* **	1/1675	2
William Payne*	11/1681	4
John Pearson* **	3/1667	1
William Perrey**	11/1678	9
Thomas Pigot*	12/1679	2
John Ray	11/1667	5
Thomas Smith	12/1677	2
Nathaniel Vincent	11/1683	7
Isaac Vossius	4/1664	1
Edward Waterhouse	7/1663	6
Sir George Wheler	12/1677	3
Robert Wood**	4/1681	4
Edward Wetenhall*	7/1683	1

SOURCES: Thomas Birch, *The History of the Royal Society*, 4 vols. (London: n.p., 1756-57; reprint ed., New York: Johnson Reprint Co., The Sources of Science, no. 44, 1968), and Gail Ewald Scala, "An Index of Proper Names in Thomas Birch, *The History of the Royal Society* (London, 1756-57)," *Notes and Records of the Royal Society of London* 28 (April 1974):263-329.

*Never corresponded with the Society

**Elected to the Council of the Society

TABLE 3.6

CLERGY IN TABLE 3.5 WHO WERE ELECTED TO THE COUNCIL OF THE ROYAL SOCIETY

Name	Year of Term	Number of Times Birch's *History* Records Attendance at Council Meetings
Arderne	1673	0
Flamsteed	1682	5
	1684	7
Gale	1679	6
	1680	6
	1681	4
	1682	2
	1684	2
	1686	11
	1687	2
Mapletoft	1677	0
	1679	4
Milles	1677	1
Pearson	1676	1
Perrey	1680	1
	1681	12
	1683	7
	1686	14
	1688	1
Wood	1682	9

SOURCES: Thomas Birch, *The History of the Royal Society*, 4 vols. (London: n.p., 1756-57; reprint ed., New York: Johnson Reprint Co., The Sources of Science, no. 44, 1968); Gail Ewald Scala, "An Index of Proper Names in Thomas Birch, *The History of the Royal Society* (London, 1756-57)," *Notes and Records of the Royal Society of London* 28 (April 1974):263-329; and the Catalogue of Fellows 1660-1685 in Michael Cyril William Hunter, "The Social Basis and Changing Fortunes of an Early Scientific Institution, 1660-1685," *Notes and Records of the Royal Society of London* 31 (July 1976):97-114.

TABLE 3.7

CLERGY ELECTED TO THE ROYAL SOCIETY AFTER THE SIGNING OF THE SECOND CHARTER GROUPED ACCORDING TO THEIR LEVEL OF ACTIVITY

Inactive	Occasional Correspondence	Sustained Correspondence	Occasional Attendance	Sustained Attendance	Long-term Interest
Burnet	Baker	Ashe	Arderne	Flamsteed	Beale
Castell	Bernard	Beale	Bathurst	Gale	Flamsteed
Clerke	Hotham	Glanvill	Horneck	Harrington	Gale
Dolben	Potter	Ray	Milles	Mapletoft	Perrey
Henchman	Rogers		Pearson	Payne	Ray
Jenkes	Sheldon		Pigot	Perrey	
Laney	Stearne		Smith	Vincent	
More	Tillotson		Vossius	Waterhouse	
Morley			Wetenhall	Wheler	
Parker				Wood	
Williams					
Woodford					
Woodroffe					
Wotton					

SOURCES: Thomas Birch, *The History of the Royal Society*, 4 vols. (London: n.p., 1756-57; reprint ed., New York: Johnson Reprint Co., The Sources of Science, no. 44, 1968) and Gail Ewald Scala, "An Index of Proper Names in Thomas Birch, *The History of the Royal Society* (London, 1756-57)," *Notes and Records of the Royal Society of London* 28 (April 1974):263-329.

the former. With this reservation, Table 3.7 can be taken to construe roughly the level of scientific accomplishment among the clerical virtuosi. There is little evidence, for example, that the majority of those listed under the 'inactive' and the 'occasional correspondence' columns did anything to advance natural philosophy. Table 3.8 shows that half the clerics elected between the issuing of the second charter and the end of 1687, were either completely inactive or infrequent correspondents. This is a sizable amount of deadwood. Most of this deadwood entered the Society in the 1660's. Two-thirds of those classed 'inactive' were elected by November 1667. Seventy-five percent of the clerics listed as occasional correspondents were elected before December 1673.

The Society appears to have been more discriminating after 1673. Eighty-six percent of those exhibiting sustained attendance were elected after 1673—the majority being elected in the period which coincided with the presidency of Sir Christopher Wren. Whereas the majority of clerics elected before 1673, showed little or no interest in the proceedings of the Society after their election, this is quite the opposite for clerics elected after 1673. At least two-thirds of those elected after 1673, attended occasionally or demonstrated sustained attendance or correspondence.

Of the 18 clerics elected between May 1663 and November 1667:

44.4%	(8 of 18)	were inactive
16.7%	(3 of 18)	corresponded occasionally
16.7%	(3 of 18)	attended occasionally
5.6%	(1 of 18)	attended regularly
16.7%	(3 of 18)	corresponded regularly

If we add the three who were unordained at the time of their election (e.g., Clerke, Waterhouse, and Woodford), our total rises to twenty-one. Nine of the twenty-one (e.g., Burnet, Clerke, Dolben, Henchman, Laney, More, Morley, Parker, and Woodford) do not appear to have either attended or corresponded with the Royal Society after their admission.

Of these nine, Clerke and More must be singled out for some mention. Both were active in scientific circles at Oxford and Cambridge.[12] We might add to this list three more (Sheldon, Stearne, and Williams) whose contributions were minimal. On the day of his admission, Williams was requested to bring in his collection.[13] Stearne left the Society a posthumous account of the weather.[14]

TABLE 3.8

THE CORRELATION BETWEEN DATE OF ELECTION AND LEVEL OF ACTIVITY OF THE FORTY CLERICAL VIRTUOSI ELECTED TO THE SOCIETY BETWEEN MAY 1663 AND DECEMBER 1687

Number	Percentage	Elected Between
	Of the 12 Who Were Inactive:	
8 out of 12	66.7%	5/1663 - 11/1667
1 out of 12	8.4%	12/1667 - 11/1673
2 out of 12	16.6%	12/1673 - 7/1678
0 out of 12	0	8/1678 - 3/1681
0 out of 12	0	4/1681 - 2/1685
1 out of 12	8.4%	3/1685 - 11/1687
	Of the 8 Who Corresponded Occasionally:	
3 out of 8	37.5%	5/1663 - 11/1667
3 out of 8	37.5%	12/1667 - 11/1673
0 out of 8	0	12/1673 - 7/1678
0 out of 8	0	8/1678 - 3/1681
0 out of 8	0	4/1681 - 2/1685
2 out of 8	25.0%	3/1685 - 11/1687
	Of the 9 Who Attended Occasionally:	
3 out of 9	33.3%	5/1663 - 11/1667
2 out of 9	22.2%	12/1667 - 11/1673
2 out of 9	22.2%	12/1673 - 7/1678
1 out of 9	11.1%	8/1678 - 3/1681
1 out of 9	11.1%	4/1681 - 2/1685
0 out of 9	0	3/1685 - 11/1687

TABLE 3.8—*Continued*

Number	Percentage	Elected Between
	Of the 4 Corresponded Regularly:	
3 out of 4	75.0%	5/1663 - 11/1667
0 out of 4	0	12/1667 - 11/1673
0 out of 4	0	12/1673 - 7/1678
0 out of 4	0	8/1678 - 3/1681
0 out of 4	0	4/1681 - 2/1685
1 out of 4	25.0%	3/1685 - 11/1687
	Of the 7 Who Attended Regularly:	
1 out of 7	14.2%	5/1663 - 11/1667
0 out of 7	0	12/1667 - 11/1673
2 out of 7	28.6%	12/1673 - 7/1678
1 out of 7	14.2%	8/1678 - 3/1681
3 out of 7	42.9%	4/1681 - 2/1685
0 out of 7	0	3/1685 - 11/1687

SOURCES: Thomas Birch, *The History of the Royal Society*, 4 vols. (London: n.p., 1756-57; reprint ed., New York: Johnson Reprint Co., The Sources of Science, no. 44, 1968); Gail Ewald Scala, "An Index of Proper Names in Thomas Birch, *The History of the Royal Society* (London, 1756-57)," *Notes and Records of the Royal Society of London* 28 (April 1974):263-329; and the Catalogue of Fellows 1660-1685 in Michael Cyril William Hunter, "The Social Basis and Changing Fortunes of an Early Scientific Institution," *Notes and Records of the Royal Society of London* 31 (July 1976):97-114.

NOTE: The 40 clerics who are the subject of this table exclude the eight clerics, styled "Original Fellows," who were elected to the Royal Society before the date of its second charter and the five other clerics who had yet to be ordained at the time of their election.

Sheldon through Robert Hooke sent in a "curiosity" for the repository.[15] Both Sheldon and Henchman were involved in negotiations concerning Chelsea College.[16] To these twelve might be added the name of Francis Potter. He wrote two letters to John Aubrey which were read before the Society. Both communications were of interest to the Society. One concerned the transfusion of blood; the other, the measuring of time by an air strainer.[17]

This leaves eight active members. Beale and Ray represent the pole of greatest activity. Harrington and Waterhouse were active, but for very short periods of time (Harrington for about two months, Waterhouse about a year). Glanvill corresponded with the Society around the end of the 1660's and the beginning of the 1670's. Down near the bottom in terms of activity are Bathurst and Pearson. Though active before the second charter, Bathurst is recorded as attending only once thereafter. Two letters of his were read at regular meetings of the Royal Society: one in 1677 and the other in 1685.[18] Pearson is recorded as attending once. Pearson occupied a seat on the Council in 1676, but he only attended one meeting.[19]

Of the six clerics elected between December 1667 and November 1673:

16.7% (1 of 6)	were inactive
50.0% (3 of 6)	corresponded occasionally
33.3% (2 of 6)	attended occasionally
0	attended regularly
0	corresponded regularly

Between the fall of Clarendon and the rise of Danby, six clerics joined the Society. One, Woodroffe, once admitted, neither attended nor corresponded—eventually being cast out in 1682. Birch's *History* has no record of Hotham either, but there is one piece of correspondence in the collection of Henry Oldenburg's letters which is addressed to him: "Inquiries concerning the Bermudas recommended to Mr. Hotham going thither March 7, 1669/70."[20]

Of the four remaining, only Bernard and Tillotson show any measure of activity. Even here this is showing charity. Neither attended, but they did correspond. Tillotson's duties as executor of Wilkins's estate and Barrow's papers involved correspondence with the Society.[21] Bernard's activity is revealed through orders of the Council. Bernard borrowed books from the Society's library in 1673 and 1677 (returning those he borrowed in 1677, ten years later in 1687). This required the approval of the Council. The Council also opportuned Bernard to take it upon himself to provide an experiment for one session of the Society. Birch's *History* gives no evidence of his having obeyed.[22] Bernard's interests were antiquarian. He carried on a correspondence with Henry Oldenburg about texts and translations of ancient mathematicians.[23] Birch's *History* records a letter read before the Society from Leibniz where the noted German mathematician inquired of Bernard the whereabouts of a certain manuscript.[24]

The two remaining clerics elected in this period, Arderne and Horneck, attended once and twice respectively. Arderne was expelled in 1685; Horneck resigned in 1682.

Of the six clerics elected between December 1673 and July 1678:

33.3%	(2 of 6)	were inactive
0		corresponded occasionally
33.3%	(2 of 6)	attended occasionally
33.3%	(2 of 6)	attended regularly
0		corresponded regularly

If we add the two who were unordained at the time of their election (e.g. Mapletoft and Wheler), the total rises to eight. Two of the eight, Castell and Jenkes, neither attended nor corresponded with the Society. Castell took up his bond in 1682; Jenkes was expelled in 1682. Wheler was active only one year after he was admitted. He attended four meetings. In 1682, he sent the Society his book concerning his travels in Greece.[25] What is of interest here is that Wheler was not ordained until 1683, long after his activity in the Society had ceased. He was expelled in 1685. Like Wheler, Smith was another travel writer.[26] Smith attended twice and corresponded once. As a fellow of Magdalen College, Oxford, Smith was active in the proceedings of the Oxford Philosophical Society.[27] His conjecture on the tides along with several observations from his travels in Greece and Asia Minor were published in the *Philosophical Transactions*.[28]

Milles, Pepys' clergyman, is recorded as attending twice. He was also elected into the Council twice. He does not appear to have been active at the general meetings of the Society much beyond a year after his admission. He was expelled in 1685.

Mapletoft was another whose period of activity in the Society occurred before he was ordained. He was Professor of physic at Gresham College from 1675 to 1679. His activity in the Society coincided with this span. He attended eight times during these years, and was elected to the Council in 1677 and 1679.

In contrast to the fellows commented upon so far, Flamsteed and Gale were extremely active. Though not elected to the Society until 1677, Flamsteed was already contributing to its proceedings in the early 1670's. His first contact with the Royal Society occurred on 13 January 1669/70 with the anonymous submission of a ephemeris on lunar occultations for the year 1670.[29] It was ordered to be printed in the *Philosophical Transactions* as were other ephemerides he submitted in 1671 and 1672.[30] Impressed by his researches, John Collins and Henry Oldenburg engaged him in correspondence. Many of the letters he wrote to Oldenburg in the early 1670's were read before the Society.[31] Through Oldenburg and Collins, Flamsteed came to the attention of Sir Jonas Moore who engineered his appointment to the office of royal astronomer in 1675.[32] Moore also played a role in Flamsteed's election to the Society in 1677.[33]

Flamsteed attended his first meeting as a fellow of the Royal Society on 13 February 1678/79. Though his attendance at its meetings began to wane during the following decade, his attendance became more frequent during the years he served on the Council in the early 1680's.[34] Flamsteed's contributions to the proceedings of the Society were largely in astronomy, but he also spoke during meetings on such varied topics as the climate at Greenwich, the specific gravity of various substances, unusual tides, and why sponges were not darker because of the light their holes absorbed.[35]

Gale was the most active of the clerics who were elected to the Society after 1663. Admitted to the Society in December 1677, Gale was elected to the Council at the end of November 1678, and became secretary of the Society in November of the following year. The frequency of his attendance at Council and regular meetings was highest in the years that he served as secretary. In Gale's first year as secretary in 1680, he attended at least 14 regular meetings of the Royal Society and 6 meetings of the Council. Gale's attendance never approached this frequency again until 1686, the first year he resumed being secretary of the Society after a three year hiatus. His hiatus coincided with his lowest record of attendance. From 1683 through 1685, Gale attended on average only 4 or 5 regular meetings of the Society per year.

As a scholar and antiquarian, Gale showed particular concern for the library of Royal Society and the state of its collections. Gale was responsible for overseeing the bequest by the Duke of Norfolk of his library to the Society. It was Gale who insisted that the Domesday book, one of the volumes in the benefaction, be returned by the herald's office to the custody of the Society.[36] Gale proposed that Norfolk's library be housed in Gresham College and he authored the inscription that declared that the library was named in honor of its benefactor.[37] Gale also composed the rules that governed the use and operations of the Bibliotheca Norfolciana.[38] He also may have played a role in appointing William Perrey its first librarian, for Gale had been Perrey's tutor when Perrey was a student at Trinity College, Cambridge.

Gale's comments at regular meetings are indicative of his belief in the knowledge to be gained from the recovery of ancient wisdom. Through his knowledge of manuscript sources, Gale persuaded the Society to take an interest in Roger Bacon as "the first, who had begun experimental philosophy."[39] On another occasion Gale used his knowledge of ancient sources to reconstruct the formula for the production of Greek fire.[40] The majority of his contributions to the proceedings of the Society were citations from ancient authors. At one meeting of the Society, Gale cited a passage from Epiphanius that described a rock very similar to the one then under discussion by the Society.[41] At another meeting of the Society, Gale cited a passage from Vitruvius to explain the use of an ancient brick that had recently been unearthed from an archaeological dig.[42]

Like many within the Society, Gale had an interest in archaeology. On 18 June 1684, the journal book records that Gale "gave an account of a great earthen urn, found lately at Peckham in Surrey."[43] When the Society expressed an interest in obtaining such an urn, Gale managed to locate for the Society "a large earthen urn, with the bones in it, and an earthen cover upon it" from the same gravel pit near Peckham of which he had spoken before. Accompanying this find was "a paper signed by several persons, who were present at the finding of this urn and seven or eight urns more, which were not so intire."[44]

Of the two clerics elected between August 1678 and March 1681:

0	were inactive
0	corresponded occasionally
50.0% (1 of 2)	attended occasionally
50.0% (1 of 2)	attended regularly
0	corresponded regularly

Perrey and Pigot stand in marked contrast. Pigot attended twice. He spoke both times—pointing to an anomaly in the theory surrounding an hermetically sealed thermometer at one meeting and presenting a strange sort of bituminous "earth" at another. Though his account of an earthquake was subsequently published in the *Philosophical Transactions*, his distance from London prevented him from being more active in the proceedings of the Society. He died in 1686.[45] Perrey on the other hand was very active. During the 1680's, he attended nine meetings and was elected to the Council in 1680, 1681, 1683, 1686, and 1688. His main responsibility was that of librarian to the Royal Society collections.

Of the six clerics elected between April 1681 and February 1685:

0	were inactive
33.3% (2 of 6)	corresponded occasionally
16.7% (1 of 6)	attended occasionally
50.0% (3 of 6)	attended regularly
0	corresponded regularly

Of the six elected, Rogers, Wetenhall, and Baker occupy the bottom rung on the ladder of activity. Rogers never attended a meeting and he never corresponded directly with the Society—Aubrey read a portion of his weather journal at a meeting on April 20, 1681, the same month as his election.[46] Wetenhall attended only one meeting about an half year after he was elected. Baker was a man of some scientific attainment, but though elected, he never traveled from county Durham to sign the charter book and be admitted. He never corresponded directly with the Society, but the Society did see to the publication of his book, *The Geometrical Key* in 1682.[47] Payne was active for little more than a year after his election. Vincent was active a bit longer, but he inexplicably resigned on January 12, 1686/87. Both made legitimate contributions—not just random observations—at the meetings they attended. Indeed, Vincent had the honor of presenting Isaac Newton's *Principia* to the Society and of reviewing Boyle's *Vulgar Notion of Nature*.[48] Wood, a compiler of almanacs, was relatively active in the early 1680's. He was a member of the Council in 1682.

Of the two clerics elected between March 1685 and November 1687:

0	were inactive
0	corresponded occasionally
0	attended occasionally

50.0% (1 of 2)	corresponded regularly
0	attended regularly

Wotton, elected in December 1687, was not active until after 1688. A member of the Dublin Philosophical Society, Ashe is mentioned frequently in its minutes.[49] Elected to the Royal Society in February 1686, he had to wait four months before he was able to journey to London to sign the charter-book.

If we look at the forty-five clerical fellows as a whole, we find an initial flurry of activity around the mid-1660's. The period of greatest activity (that is, the period where instances of correspondence and attendance are most frequent) begins about 1675, building to a crescendo in 1678-79, and peaking in 1681-84. Given Dorothy Stimson's contention that the Society experienced a rejuvenation under the presidency of Wren, this is what might be expected.[50]

NOTES

1 This figure is compiled from a count of the names in the Chronological Register of Fellows in Royal Society of London, *The Record of The Royal Society of London*, 3rd. ed. (London: Oxford University Press, 1912), pp. 375-516. Michael Hunter declares that the register represents "the most accurate list of Fellows" that can presently be obtained. See Michael Hunter, "The Social Basis and Changing Fortunes of an Early Scientific Institution: An Analysis of the Membership of the Royal Society, 1660-1685," *Notes and Records of the Royal Society of London* 31 (July 1976):44.

2 In addition to the forty-eight, another five became ordained after the date of their election. They were Henry Clerke, John Mapletoft, Edward Waterhouse, Sir George Wheler, and Samuel Woodford.

3 Hunter, "Membership," p. 37, Table 5.

4 See Table 2 in Hunter, "Membership," p. 26. Not every fellow elected appears in the printed lists of members that the Society produced. These lists omit many who failed to undergo the formal process of being admitted to the Royal Society and of signing their names in its Charter Book. See Hunter, "Membership," pp. 44-46, for a discussion of the accuracy of these lists.

5 Thomas Birch, *The History of the Royal Society of London for Improving of Natural Knowledge From its First Rise. In which The most considerable of those Papers communicated to the Society, which have hitherto not been published, are inserted in their proper order, as a supplement to The Philosophical Transactions*, with an introduction by A. Rupert Hall and additional bibliographical note by Marie Boas Hall, 4 vols. (London: n.p., 1756-57; reprint ed., New York: Johnson Reprint Co., The Sources of Science, no. 44, 1968).

6 Gail Ewald Scala, "An Index of Proper Names in Thomas Birch, *The History of the Royal Society* (London, 1756-1757)," *Notes and Records of the Royal Society of London* 28 (April 1974):263-329.

7 See Hunter's discussion of this problem in Hunter, "Membership," pp. 47-49.

8 Sir Henry Lyons, *The Royal Society 1660-1940* (Cambridge: Cambridge University Press, 1944), p. 72, and Dorothy Stimson, *Scientists and Amateurs: A History of the Royal Society* (New York: Henry Schuman, Inc., 1948; reprint ed., New York: Greenwood Press, 1968), pp. 98-99.

9 Hunter, "Membership," p. 49.

10 Data on Williams, Burnet, Tillotson, and Castell regarding the payment of their dues compiled from the Catalogue of Fellows 1660-1685 in Hunter, "Membership," pp. 97-114.

11 John Mapletoft was not present at a meeting of the Council on 7 August 1679, yet the minutes record that "it was ordered,... that Sir John Hoskyns, Sir Jonas Moore, Mr. Colwall, Mr. Hill, Dr. Gale, Dr. Brown, Dr. Grew, Dr. Mapletoft, or any three of the council, do meet together, and go to Mr. Cheney to discourse with him concerning Chelsea College:..." Birch, *History*, 3:501.

12 A physician and minor scientist with an interest in anatomy, Henry Clerke was friendly with Ralph Bathurst, Richard Lower, Thomas Millington, William Petty, and Thomas Willis. See Robert Gregg Frank, Jr., *Harvey and the Oxford Physiologists: A Study of Scientific Ideas and Social Interaction* (Berkeley: University of California Press, 1980), p. 70.

For nearly fifty years, the Cambridge Platonist, Henry More, was a fellow and a tutor at Christ's College, Cambridge. Joseph Glanvill and Robert Boyle were among his friends and correspondents. His works influenced many during their formative years at Cambridge. For the impact of More's Cartesianism upon the thought of Henry Power, see Charles Webster, "Henry Power's Experimental Philosophy," *Ambix* 14 (1967):150-78. More's *Antidote Against Atheism* (1653) shaped John Ray's subsequent formulation of the argument from design. See William H. Austin, "Henry More," *Dictionary of Scientific Biography*, ed. Charles Coulston Gillispie, 16 vols. (New York: Charles Scribner's Sons, 1970-80), 9:509 (hereafter cited as *DSB*). Newton's debt to More is difficult to gauge. See the comments of Richard S. Westfall in his *Never at Rest: A Biography of Isaac Newton* (Cambridge: Cambridge University Press, 1980), pp. 97, 301, 304. E. A. Burtt, *The Metaphysical Foundations of Modern Science* (Garden City, NY: Doubleday & Co., Anchor Books, 1954), pp. 135-148, and Alexander Koyre, *From the Closed World to the Infinite Universe* (Baltimore: The Johns Hopkins Press, 1957), chap. 6, suggest that More's metaphysics was instrumental in the formation of Newton's concepts of space and time. J. E. McGuire sees a connection between More's neoplatonism and Newton's conception of force. See J. E. McGuire and P. M. Rattansi, "Newton and the 'Pipes of Pan'," *Notes and Records of the Royal Society of London* 21 (1966):108-143; J. E. McGuire, "Force, Active Principles, and Newton's Invisible Realm," *Ambix* 15 (1968):154-208; and J. E. McGuire, "Neoplatonism and Active Principles: Newton and the Corpus Hermeticum," in *Hermeticism and the Scientific Revolution*, ed. Robert S. Westman and J. E. McGuire (Los Angeles: William Andrews Clark Memorial Library, University of California, 1977).

During the 1650's, More was Cambridge's foremost apostle of the philosophy of Descartes. More corresponded with Descartes from 1648-49, and promoted the study of his works until the late 1660's. See Charles Webster, "Henry More and Descartes: Some New Sources," *The British Journal for the History of Science* 4 (December 1969):359-77. In the late 1660's, More abandoned Cartesianism as he came to see that its mechanistic system banished God from the universe.

More's public disavowal of mechanical philosophy and his use of Boyle's experiments to bolster a metaphysics which was anti-mechanistic helped to foster the impression that More was critical of the Society and the philosophy it professed. For the reaction of one contemporary, see John Beale to Henry Oldenburg, 24 June 1671, *The Correspondence of Henry Oldenburg*, trans. and ed. A. Rupert Hall and Marie Boas Hall, 13 vols. (vols. 1-9: Madison: The University of Wisconsin Press, 1965-75; vols. 10-11: London: Mansell, 1975-77; vol. 13: London: Taylor & Francis, 1986), 8:119-20. For Boyle's response to More's attempt to reinterpret his experiments, see Robert Boyle, *Tracts written by the Honourable Robert Boyle, containing New Experiments, touching the relation betwixt flame and air. And about explosions. An hydrostatical discourse occasion'd by some objections of Dr. Henry More against some explications of new experiments made by the author of these tracts: To which is annex't, An hydrostatical letter dilucidating an experiment about the way of weighing water in water. New experiments, Of the positive or relative levity of bodies under water. Of the air's spring on bodies under water. About the differing pressure of heavy solids and fluids* (London: Printed for Richard Davis, bookseller in Oxon, 1672), and R. A. Greene, "Henry More and Robert Boyle on the spirit of nature," *Journal of the History of Ideas* 23 (October-December 1962):451-74.

When Henry Stubbe numbered More among the opponents to the Society, More responded with a letter which was published in Joseph Glanvill's *A Praefatory Answer to Mr. Henry Stubbe, The Doctor of Warwick. Wherein the Malignity of his Temper, the Hypocrisie of his Pretences, the Falshood of his Reports, And the Impertinency of his Arguings & Quotations in his Animadversions on Plus Ultra, Are discovered* (London: Printed by

A. Clark, for J. Collins, at the Kings Arms in Ludgate-street, near the West-end of S. Pauls; and at the Kings Head in Westminster-hall, 1671), pp. 154-58. More attacked Stubbe for misconstruing his position. He affirmed there was a difference between the experimental philosophy which the Society pursued and the "*Mechanical Philosophy*" that "inclines men to *Atheism*."

13 Birch, *History*, 1:388.

14 Birch, *History*, 4:291.

15 Birch, *History*, 2:462. According to the minutes of the meeting on 12 January 1670/71, the curiosity sent in by Sheldon was presumed to be "several pieces of an hippopotamus" dug up at Chatham.

16 See Birch, *History*, 2:77, 79, 238. In April 1669, Charles II granted the Society the buildings and grounds of Chelsea College. Founded in the reign of James I, it had long since ceased to function as a Protestant seminary. Though the Society spent time and money clearing its title to the property, the buildings and grounds were too decayed to return an adequate income. In 1682, the King purchased the property for 1300 pounds and built a military hospital. The Society invested the proceeds in East India Company stock. For a history of the Society's troubles concerning Chelsea College, see R. K. Bluhm, "Remarks on the Royal Society's Finances," *Notes and Records of the Royal Society of London* 13 (November 1958):98; Michael Cyril William Hunter, *Science and Society in Restoration England* (Cambridge: Cambridge University Press, 1981), pp. 40, 127-28; and Lyons, *Society*, 93.

17 For Potter's letter concerning blood transfusions, see Birch, *History*, 2:361. On Potter's use of an air strainer to measure time, see Birch, *History*, 2:224.

18 After 1663, Bathurst's only recorded attendance at a regular meeting of the Society occurred on 29 June 1664. See Birch, *History*, 1:444. Bathurst was instrumental in forwarding to the Society an account by Roger Moslyn of "unusual damps, and their odd effects" in a coal mine in Flintshire. This was read before the Society on 10 May 1677, and subsequently published in the *Philosophical Transactions*. See Birch, *History*, 3:339, and Roger Moslyn, "A Relation of some strange phaenomena, accompanied with mischievous effects in a Cole work in Flintshire; sent March 31. 1677. to the Reverend and eminently Learned Dr. Bathurst, Dean of Bath and Wells, by an Ingenious Gentleman, Mr. Roger Moslyn, of the Inner Temple, who, at the said Doctor's request obtained it from his Fathers Steward, and Overseer of his Cole-works, who was upon the place when the thing was done; the same Mr. Moslyn being also assured of it from his Father, Sr. Roger Moslyn, Lord of the Mannor, and several others, who were Eye-witnesses," *Philosophical Transactions* 12 (25 June 1677):895-99. At another meeting of the Society on 11 November 1685, William Musgrave read a letter from Dr. Bathurst . See Birch, *History*, 4:429. One further piece of correspondence is recorded in the minutes. On the 22 March 1681/82, Bathurst enclosed 10 pounds in a letter to the Society. See Birch, *History*, 4:137.

19 On 18 July 1678, the minutes record that Pearson spoke at a regular meeting of the Society. See Birch, *History*, 3:425. His only attendance at a Council meeting occurred on the 6 March 1675/76. See Birch, *History*, 3:309.

20 See Henry Oldenburg to Charles Hotham, 7 March 1669/70, in Oldenburg, *Correspondence*, 6:535-37. Charles Howard (d. 1713) cited Hotham in a report that was read before the Royal Society on 18 February 1662/63, concerning "sulpherous vapors issuing out of the earth near Wiggan in Lancashire." Howard wrote:

> "'I was informed, that Mr. Hotham, who was then minister of Wiggan, hearing of this, went to the place, and brought away some water from thence with him, and having distilled it, he found, that the vapour had not communicated any of its

quality to the water. The vapour was visible to me till it was set on fire, with little or no scent; the colour of the flame no much differing from aquavitae, but much hotter. I am confident it would serve very well to boil meat, and might be very proper for many chemical operations.'"

See Birch, *History*, 1:199. This is the only mention of Hotham's name in Birch's *History* beyond the routine recording of his nomination, election and admission in January 1667/68. Hotham is the only dissenter among the clerical virtuosi. He was ejected from his living at Wigan in 1662. See Oldenburg, *Correspondence*, 6:536n.

21 As executor of Wilkins's estate, Tillotson honored Wilkins's request that the Society receive 400 pounds from his estate. See Birch, *History*, 3:93, 110-11. On 25 July 1683, the minutes record that "Mr. Abraham Hill, in his own name and that of Dr. Tillotson, dean of Canterbury, presented to the Society for their Library a manuscript of Dr. Isaac Barrow, containing his demonstrations of Archimedes and Apollonius." See Birch, *History*, 4:217. Tillotson was also instrumental in conveying to the Society a paper concerning comets by Thomas Hill. See Birch, *History*, 4:66.

22 Concerning the books the Society lent Bernard, see Birch, *History*, 3:129, 3:342, 4:548. For the Society's request to Bernard to provide it with an experiment, see Birch, *History*, 3:160.

23 See Henry Oldenburg to Edward Bernard, 13 September 1670, in Oldenburg, *Correspondence*, 7:159-60; Bernard to Oldenburg, 24 November 1670, 7:287-88; Oldenburg to Bernard, 1 December 1670, 7:292-93; Oldenburg to Bernard, 27 June 1671, 8:126-28; Oldenburg to Bernard, 4 February 1671/72, 8:523-24; Bernard to Oldenburg, 14 September 1672, 9:242-43; Bernard to Oldenburg, 15 April 1673, 9:592-93; and Bernard to Oldenburg, 28 April 1673, 9:635-36.

24 Gottfried Wilhelm Leibniz to Robert Hooke, 8 April 1680, in Birch, *History*, 4:33.

25 Birch, *History*, 4:171. The book presumably was *A Journey into Greece By George Wheler Esq; In company of Dr Spon of Lyons. In Six Books. Containing I. A Voyage from Venice to Constantinople. II. An Account of Constantinople and the Adjacent Places. III. A Voyage through Lesser Asia. IV. A Voyage from Zant through several Parts of Greece to Athens. V. An Account of Athens. VI. Several Journeys from Athens, into Attica, Corinth, Boetia, &c. With variety of Sculptures* (London: Printed for William Cademan, Robert Kettlewell, and Awnsham Churchill, at the Popes Head in the New Exchange, the Hand and Scepter in Fleetstreet, and the Black Swan near Amen-Corner, 1682).

26 See Tho[mas] Smith, *Remarks Upon the Manners, Religion and Government Of the Turks. Together with A Survey of the Seven Churches of Asia, As they now lye in their Ruines: and A Brief Description of Constantinople* (London: Printed for Moses Pitt, at the Angel in St. Pauls Church-yard, 1678) and Tho[mas] Smith, *An Account of the Greek Church, as to Its Doctrine and Rites of Worship: With Several Historicall Remarks interspersed, relating thereunto. To which is added An Account of the State of the Greek Church, under Cyrillus Lucaris Patriarch of Constantinople, with a Relation of his Sufferings and Death* (London: Printed by Miles Flesher for Richard Davis in Oxford, 1680).

27 Smith's name appears along with those of Bathurst, Bernard, Pigot, and Wallis in the list drawn up by Robert Plot of the founding members of the Oxford Philosophical Society. See Anthony à Wood, *Athenae Oxonienses. An Exact History of All the Writers and Bishops Who Have Had Their Education in the University of Oxford. To Which Are Added, The Fasti, or Annals of the Said University*, ed. Philip Bliss, 3rd. ed., 4 vols. (London: Printed for F. C. and J. Rivington; Lackington, Allen, and Co.; T. Payne; White, Cochrane, and Co.; Longman, Hurst, Rees, Orme, and Brown; Cadell and Davies; J. and A. Arch; J. Mawman; Black, Parry, and Co.; R. H. Evans; J. Booth, London; and J. Parker, Oxford, 1813-20; facs. reprint ed., Hildesheim: George Olms, Anglistica & Americana, no. 22, 1969), 1:xcviii.

28 See Thomas Smith, "Historical Observations relating to Constantinople," *Philosophical Transactions* 13 (20 October 1683):335-46; "An Account of the City of Prusa in Bithynia, and a continuation of the Historical Observations relating to Constantinople," *Philosophical Transactions* 14 (20 January 1683/84):431-54; "A conjecture about an under-Current at the Streights-Mouth read before the Oxford Society, Dec. 21. 1683," *Philosophical Transactions* 14 (20 April 1684):564-66; and "A Journal of a Voyage from England to Constantinople, made in the Year 1668 by T. Smith D.D. & F.R.S.," *Philosophical Transactions* 19 (July 1697):597-619.

29 Birch, *History*, 2:415.

30 See John Flamsteed, "An Accompt of such of the more notable Celestial Appearances of the Year 1670, as will be conspicuous in the English horizon; written by the Learned and Industrious Mr. John Flamstead Novemb. 4 1669. and by him addressed and recommended for encouragement, to the Right Honorable, the Lord Viscount Brouncker, as President of the Royal Society," *Philosophical Transactions* 4 (17 January 1670):1099-1112. Other ephemerides of Flamsteed were presented to the Society on 24 November 1670, 23 November 1671, and 20 November 1672. See Birch, *History*, 2:454, 2:495, 3:61. They too were printed in the *Philosophical Transactions*. See John Flamsteed, "Lunae ad Fixas Saturnumque, Anno 1671. Appulses observabiles, praedicti, & ad Meridianum Latitudinemque Londini accurate supputati a Joh. Flamsteadio Derbiensi-Anglo," *Philosophical Transactions* 5 (12 December 1670):2029-34; John Flamsteed, "Lunae ad Fixas Appulsus Visibiles, nec non arctiores juxta eas Transitus, observabiles A. 1672: Praedicti, & ad Meridianum Latitudinemque Londini, e Tabulis Carolinis accurate supputati a Joh. Flamstedio Derbiensi-Anglo," *Philosophical Transactions* 6 (20 November 1671):2297-3001; and John Flamsteed, "An Extract of Mr. Flamstead's Letter written to the Publisher from Derby Novemb 16. 1672. concerning the Appulses, by him calculated for A. 1673. of the Moon, and the other Planets to Fixed Stars; together with an Observation of the Planet Mars, made by the same," *Philosophical Transactions* 7 (16 December 1672):5118-24.

On 30 October 1672, Flamsteed's tract on the equation of time was read before the Society. See Birch, *History*, 3:58. This tract together with Flamsteed's emendations of Horrock's lunar theory was published by John Wallis in the posthumous edition of Horrocks's works. See Jeremiah Horrocks, *Jeremiae Horroccii, Liverpoliensis Angli, ex Platinatu Lancastriae Opera Posthuma; viz. Astronomia Kepleriana, defensa & promota. Excerpta ex Epistolis ad Crabtraeum suum. Observationum Coelestium Catalogus. Lunae Theoria nova. Accedunt Guilielmi Crabtraei, Mancestriensis, Observationes Coelestes. In calce adjiciuntur Johannis Flamstedii, Derbiensis, De Temporis Aequatione Diatriba. Numeri ad Lunae Theoriam Horroccianam*, ed. John Wallis (London: Typis Gulielmi Godbid, Impensis J. Martyn Regalis Societatis Typographi, ad insigne Campanae in Coemeterio D. Pauli, 1673).

31 Eight letters from Flamsteed to Oldenburg were read before the Society. See Birch, *History*, 2:424, 2:454, 2:482, 3:9, 3:10, 3:61, 3:82, 3:339. For the text of Flamsteed's letter, 7 February 1669/70, which was read before the Society on 17 February 1669/70, see Oldenburg, *Correspondence*, 6:468-70. For the text of Flamsteed's letter, 16 November 1670, which was read before the Society on 24 November 1670, see Oldenburg, *Correspondence*, 7:265-67. This was subsequently published in the *Philosophical Transactions*. See Flamsteed, "Lunae, Anno 1671," pp. 2029-34. For the text of Flamsteed's letter, 13 May 1671, which was read before the Society on 18 May 1671, see Oldenburg, *Correspondence*, 8:45-48. For the text of Flamsteed's letter, 23 December 1671, which was read before the Society on 1 February 1671/72, see Oldenburg, *Correspondence*, 8:427-28. This too was published in the *Philosophical Transactions*. See "De Lunae ad Pleiades reliquas Applicationibus, Anno 1672 observandis, promissa appendicula Johannis Flamsteed," *Philosophical Transactions* 6 (22 January 1671/72):3061-63. For the text of Flamsteed's letter, 5 February 1671/72, which was read before the Society on 8 February

1671/72, see Oldenburg, *Correspondence*, 8:524-26. For the text of Flamsteed's letter, 16 November 1672, which was read before the Society on 20 November 1672, see Oldenburg, *Correspondence*, 9:326-28. It too was published in the *Philosophical Transactions*. See Flamsteed, "Extract," pp. 5118-24. For the text of Flamsteed's letter, 17 February 1672/73, which was read before the Society on 2 April 1673, see Oldenburg, *Correspondence*, 9:468-70. For the text of Flamsteed's letter, [?] May 1677, which was read before the Society on 17 May 1677, see Oldenburg, *Correspondence*, 13:283-85. It too was printed in the *Philosophical Transactions*. See "Mr. Flamstead's account of his Observations of the late Comet, sent in a letter to the Publisher, Greenwich, May 18. 1677," *Philosophical Transactions* 12 (26 May 1677):873-75.

32 See Victor E. Thoren, "John Flamsteed," in *DSB*, 5:22-23.

33 See the minutes for 8 February 1676/77 in Birch, *History*, 3:333.

34 Flamsteed, for example, attended three meetings of the Society between the date he was admitted and the end of 1679. See Birch, *History*, 3:462, 495, 507-08, 511-13. Only two instances of attendance appear in Birch's *History* for 1680 and 1681. See Birch, *History*, 4:17-19, 100. His attendance picked up in 1682 when he was elected to the Council, trailed off somewhat in 1683 when he was dropped from the Council, and picked up again in 1684 when the Society reelected him to a seat on the Council. In 1684, he attended 8 Council and 6 regular meetings of the Society.

35 For Flamsteed's comments concerning Greenwich, see Birch, *History*, 3:495. For his observations concerning specific gravity, see Birch, *History*, 3:511-13. For his account of unusual tides and rock formations, see Birch, *History*, 4:288 and 4:175.

36 Birch, *History*, 3:440, 4:79, 4:90.

37 Gale's proposal is contained in the rules he drew up for "the government of the Bibliotheca Norfolciana." See Birch, *History*, 3:451. For the role Gale played in producing the inscription, see Birch, *History*, 4:15, 26.

38 Gale's "Orders concerning the government of the Bibliotheca Norfolciana" is published in Birch, *History*, 3:451. His concern for the library also extended to a concern for the preservation of the Society's papers. On 10 December 1679, Gale was appointed to a committee "to view the former Letter-books, to see how perfect the entries are, and to take an account of all loose letters recovered from the administratix of Mr. Oldenburg, and of those, that have come to the secretaries since; and to take care, that those, that are not yet entered, be entered forthwith, and that the originals be with all care got together, in order to their being preserved in such a way as the Society shall direct." For Gale's appointment to this committee, see Birch, *History*, 3:515, 518.

39 Birch, *History*, 3:479. See also Birch, *History*, 3:470-71, 3:472, 3:473-74, 4:298.

40 Birch, *History*, 4:296, 348.

41 Birch, *History*, 4:114.

42 Birch, *History*, 4:461.

43 Birch, *History*, 4:303.

44 Birch, *History*, 4:425-26. Gale presented the urn to a meeting of the Society on 28 October 1685.

45 For Pigot's criticism of a theory regarding an hermetically sealed thermometer, see Birch, *History*, 4:297-98. Pigot's presentation of the unusual rock sample is noted in Birch,

History, 4:484. For Pigot's report on the earthquake that was felt in Oxford, see Tho[mas] Pigot, "An Account of the Earthquake that happened at Oxford and the parts adjacent Sept. 17. 1683. by a fellow of a College in that University, and of the Royal Society," *Philosophical Transactions* 13 (20 September 1683):311-21. Pigot apparently had an interest in music theory. Wallis noted in the *Philosophical Transactions* that Pigot mistakenly took the credit that belonged to William Noble for a discovery in music. See John Wallis, "Dr. Wallis's Letter to the Publisher, concerning a new Musical Discovery; written from Oxford, March 14. 1676/77," *Philosophical Transactions* 12 (23 April 1677):840.

46 See Birch, *History*, 4:81.

47 Tho[mas] Baker, *The Geometrical Key: or the Gate of Equations Unlock'd: A New discovery of the Construction of all Equations, howsoever affected, not exceeding the fourth Degree; viz. of, Linears, Quadratics, Cubics, Biquadratics; And the finding of all their Roots, as well false, as true; without the use of Mesolabe, Trisection of Angles; without Reduction, Depression, or any other previous preparation of Equations, by a Circle, and any (and that but one only) Parabole* (London: Printed by J. Playford, for R. Clavel, at the Peacock in S. Paul's Church-yard, 1684). For the circumstances surrounding its publication, see Birch, *History*, 4:155-56, 162, 292.

48 For Vincent's presentation of I[saac] S. Newton, *Philosophiae Naturalis Principia Mathematica* (London: Jussu Societatis Regiae ac Typis Josephi Streater, 1687), see Birch, *History*, 4:479. Concerning Vincent's review of R[obert] B[oyle], *A Free Enquiry Into the Vulgarly Receiv'd Notion of Nature Made in an Essay, Address'd to a Friend* (London: Printed by H. Clark, for John Taylor at the Globe in St. Paul's Church-yard, 1685/86), see Birch, *History*, 4:487.

49 For references to Ashe in the minutes of the Dublin Philosophical Society, see Birch, *History*, 4:248, 303-04, 321-24, 342-43, 352, 375, 409-11, 416-17, and 432-33. Ashe signed the charter-book on 2 June 1686. See Birch, *History*, 4:487.

50 Stimson, *Scientists*, 99. See also Lyons, *Society*, 92-93.

– 4 –

THE PROBLEM OF "BELIEF"

An important element in the rise of modern science was its success in undermining the deference paid authorities in matters of belief. A readiness to believe what credible authorities attested had long been held a virtue before the rise of modern science. With the rise of modern science such credulity became a vice.

Rather than rely upon authorities as a source for knowledge, the new science placed its trust in observation and experiment. It resorted to authorities when it lacked other means of information, but it refused to accept their testimony if the things which they attested appeared improbable or unlikely. It withheld its assent until observation and experiment could corroborate their testimony.

This mistrust of authorities was an integral part of the ideology of science which the Society espoused. One of the themes which Sprat returns to throughout his *History of the Royal Society* was the Society's opposition to a state of mind which uncritically accepted the trustworthiness of testimony from credible authorities. Such reverence for their testimony ignored the fact that even the most credible authorities were fallible human beings. Sprat assured his readers that if the Society appeared immodest in neglecting the guidance of "wiser, and more discerning Men, then our selves," it was no disparagement of these men's character or ability, but a sober recognition "that we our selves may err, and all mankind besides."[1]

Sprat even suggested that the Society paid authorities the sincerest form of flattery when it subjected their opinions to critical examination. To question every assertion, taking nothing upon faith, was to imitate the practice credible authorities had always followed when assessing other authors. To illustrate his point, Sprat cited the ancient Greeks who purportedly "fetch'd" their learning from the "*Priests* of *Isis* and *Osiris*." Sprat maintained that had the Greeks blindly assented to all the things the priests had taught, their philosophy would have exhibited "all the infinit Idolatries, which their Masters embrac'd."[2]

Sprat asked his readers why the Society should not follow the example of the Greeks and proclaim its freedom "to distinguish, and choose, what we will follow."[3] Did not the Greeks "trust themselves and their own Reasons?" Did not the Greeks "busie themselves in inquiry, make new Arts, establish new Tenents, overthrow the old, and order all things as they pleas'd, without any servile Regard to their Predecessors?"[4] Why should the Society be denied the same liberty of judgment? Given the stark reality of human fallibility, any restriction of this liberty was a sin against the truth.

Sprat accused authorities of conspiring against this liberty when they encouraged veneration of their persons and their teachings. Such veneration helped to perpetuate their errors by suppressing free discussion of their theories and opinions. Sprat condemned their willingness to be ignorant of the truth if discovery of the truth could hurt their reputations. He blamed their pride for corrupting human knowledge and he traced its ill effects back to the birth of human learning.

The priests of ancient Egypt were the first in Sprat's opinion to let their pride obstruct the progress of human knowledge. Fearing exposure of their learning to the light of free discussion, they chose "to wrap up their Observations on Nature, and the Manners of Men, in the dark Shadows of *Hieroglyphicks*; and to conceal them, as sacred *Mysteries*, from the apprehension of the vulgar." Such a stratagem, Sprat contended, "was a sure way to beget Reverence in the Peoples Hearts towards *themselves*: but not to advance the true Philosophy of *Nature*."[5]

The same pride betrayed the Greeks when their desire for recognition overcame their love of truth. Sprat observed that in their race to gain priority for the results of their research the Greeks "fix'd, and determined their judgments, on general conclusions too soon, and so could not afterwards alter them, by any new appearances, which might represent themselves."[6] Having staked their reputations upon the soundness of their conclusions, they were loath to admit that they might have been mistaken. To secure their reputations against the discovery of their errors they declared themselves "the onely Masters of Reason" and insisted that their pupils become disciples of their teachings.[7]

Sprat maintained that it was only in his day that modern authors had found the courage to liberate themselves from their tutelage to the Greeks. Yet the sway of the Greeks could still be felt in the way some modern authors imposed "new Theories on Mens Reasons."[8] Sprat reported that their manner was so imperious that it was difficult to believe that they did not seek a tyranny over the course of human learning that was every bit as onerous as the one the Greeks imposed. Sprat thought their manner could not be squared with the liberty of judgment true philosophy embraced. The "true Philosophy of *Nature*" respected the liberty of others to criticize its teachings and it welcomed their corrections provided the evidence they presented was "establish'd on solid works, and not onely on *prejudices*, or *suspicions*."[9]

Sprat affirmed that this philosophy informed the conduct of the Society. Sprat wrote:

> To the *Royal Society* it will be at any time almost as acceptable, to be *confuted*, as to *discover*: seeing, by this means, they will accomplish their main *Design*: others will be inflam'd: many more will labour; and so the *Truth* will be obtain'd between them:...[10]

Sprat expressed surprise that a group so open to reason and correction should have counted among its opponents so many reputed men of learning. The reaction of men of learning should have come as no surprise for it was consistent with his indictment of their behavior in the past. Sprat attributed their opposition to their fears that the Society and its "Philosophy of *Nature*" would undermine respect for the value of their learning. To allay their fears, Sprat cited the men of learning who were fellows of the Society. Had the Society been indifferent to the substance of their learning, few men of learning would have bothered to have joined. They joined the Society because the Society had a need for their facility in languages and their command of ancient sources.

Sprat assured his readers that the Society was not averse to the study of ancient authors. The writings of the ancients were a source of information about the natural world and such a source could not be mined without the knowledge and expertise men of learning could provide. Their election to the Society showed the value that was placed upon this source of information.

Yet Sprat was forced to acknowledge that the Society did not handle this source of information with the requisite respect some men of learning thought was proper. Although the Society professed respect for the writings of the ancients, it did not venerate their testimony to the exclusion of other evidence. Sprat wrote:

> We approach the Antients, as we behold their Tombs, with veneration: but we would not therefore be confin'd to live in them altogether: nor would (I believe) any of those, who profess to be most addicted to their Memories.[11]

Given the fallibility of the ancients, Sprat had trouble understanding why men of learning should rank their testimony above experiment and observation as a reliable source of knowledge. To fawn upon the ancients when their testimony contradicted observation and experiment was to slight the reputation of "the most antient Author of all others, even *Nature* it self."[12] Sprat invited men of learning to make amends to Mother Nature by devoting more attention to the great volume of creation. The knowledge they would gain from a perusal of its pages would more than compensate them for the questions it would raise about the certainty of the knowledge they received from other authors.

Sprat was willing to trust the knowledge he received from other authors if this knowledge did not conflict with the great volume of creation. But there were others within the Society whose faith in other authors did not extend as far as this. They claimed to trust no other author except Nature herself. Thus, the philosopher, John Locke, thought it "idle and useless to make it one's business to study what have been other men's sentiments in things where reason is only to be judge."[13] Locke wrote:

> Truth needs no recommendation, and error is not mended by it; and in our inquiry after knowledge it as little concerns us what other men have thought as it does one who is to go from Oxford to London to know what scholars walked gently on foot, enquiring the way and surveying the country as they went, who rode past after their guide without minding the way he led, who were carried along muffled up in a coach with their company, or where an Doctor lost or went out of his way, or where a Master [of Arts] stuck in the mire. If a traveler gets a knowledge of the right way, 'tis no matter whether he knows the infinite windings, by-ways, and turnings where others have been misled; the knowledge of the right secures him from the wrong, and that is his great business; and so methinks it is in our pilgrimage through this world.[14]

Locke was convinced that a knowledge of other authors could bias a person's judgment of the great volume of creation. No one could study other authors and come away unscathed, for a mere acquaintance with their opinions could "insensibly" engage a person to post themselves in their defense. A person so engaged could not approach the book of Nature without reading the words of other authors into the meaning of its text.[15]

Locke urged his readers to be as indifferent to other authors as Nature was herself. Only by keeping "a perfect indifferency for all opinions" could they read the book of Nature and derive from its text "a true and clear notion of things as they are in themselves."[16] Locke thought a free and open mind the best preparation for mastering the truth in any subject. He cited the art of medicine as a case in point. Locke wrote:

> For example, were it my business to understand physic, would not the safer and readier way be to consult nature herself and inform myself in the history of diseases and their cures, than, espousing the principles of the dogmatists, methodists, or chemists, engage in all the disputes concerning either of those systems and suppose it true till I have tried what they can say to beat me out of it? Or supposing that Hippocrates or any other book infallibly contains the whole art of physic, would not the direct way be to study, read and consider that book, weigh and compare the parts of it to find the truth, rather than espouse the doctrines of any party, who, though they acknowledge his authority, have already interpreted and wiredrawn all his text to their own sense—the tincture whereof when I have imbibed, I am more in danger to misunderstand his true meaning than if I had come to him with a mind unprepossessed by doctors and commentators of my sect, whose reasonings, interpretation and language, which I have been used to, will of course make all chime that way and make another, and perhaps the genuine, meaning of the author seem harsh, strained and uncouth to me.[17]

Though Locke put no stock in the opinions of other authors, he recognized the fact that their knowledge of the book of Nature could supplement his own. Their works recorded facts, particularly those of past events, which could not be apprehended save by virtue of their testimony. Experiment and observation could not tell the natural philosopher whether iron placed in water had always sunk in the past. Only the reports of other authors and the record of their experiments could assure the natural philosopher that the behavior of iron in water was not a chance event.[18]

Yet Locke was quick to remind his readers that facts owed their evidence to their inclusion in the book of Nature, not to the references made to them in the works of learned authors. No fact became more credible because it was cited by learned authors unless those authors had firsthand knowledge that the fact was true. A fact whose only evidence was the testimony of a single witness had to be accepted or rejected upon that evidence alone. Its citation by learned authors did not increase its probability.[19]

Thus, neither Locke nor Sprat was willing to trust authorities without first examining their testimony. The mistrust they showed was typical of the ideology of science which the Society espoused. Their example also illustrates that different groups within the Society expressed different levels of mistrust. There were some like Thomas Sprat who would only believe authorities upon the strength of their word alone if he had no other evidence for the truth or falsehood of their testimony.

Then there were others like John Locke who declined to believe authorities upon the strength of their word alone. He demanded that their testimony bear analogy to his experience or the experience of other men. No authority, however credible, could persuade him to assent if the things which they attested appeared improbable to him.

Finally there were others like John Pearson whose definition of belief forms one of the subjects of this chapter. Pearson's mistrust was less extreme than that of Sprat or Locke. He was willing to believe authorities upon the strength of their word alone if their testimony did not conflict with truths of fact or truths of reason. Once he judged a person credible, he was was willing to believe them regardless of whether their testimony was probable or not.

Pearson conceded that most authorities were not worthy of his trust. Most authorities were fallible. They often made mistakes. None could claim to be so credible that it was safe to believe their testimony if the things which they attested appeared improbable or unlikely.

Pearson's skepticism of authorities was grounded upon experience. Experience had taught him that authorities could not be trusted to always tell the truth. What is absent from Pearson's skepticism is the element of radical doubt. It is the absence of this doubt which distinguishes Pearson's skepticism from the skepticism of Sprat and Locke. Only the latter actually questioned whether authorities could have knowledge which was absolutely certain. They denied there was any criterion which authorities could use that would allow them to determine whether the things they claimed to know were absolutely certain. Both Sprat and Locke agreed that the absence of such a criterion made it impossible for authorities to guarantee their testimony against the possibility of error.

There were precautions they could take to reduce the risk of error, but there was nothing they could do to eliminate the risk entirely. No matter how evident a truth might be, there was always the possibility that the truth might prove false. Both Sprat and Locke concluded that if authorities could not be certain about the things they claimed to know, then it was foolish to trust authorities beyond the probability of their testimony.

In contrast to Sprat and Locke, Pearson was able to conceive of cases where it was reasonable to trust authorities beyond the probability of their testimony. This difference was reflected in the way he defined belief.

Pearson defined the word, "belief," as if its meaning only extended to a particular class of beliefs. This class did not contain every assent of the mind, but only those assents which the mind gave to testimony. Though testimony furnished every object of belief, not every assent to testimony was a species of belief. Some assents to testimony were species of "opinion." What set belief apart as a species of assent was its reason, cause, or motive. No object given through testimony was an object of belief if the reason for its assent were not the credibility of the person who attested to its truth. Assents which took their motives from the probability of their objects were not species of belief but species of opinion. Beliefs took their motives from the credibility of the persons who attested to their objects. Thus, belief was something more than an assent to testimony. It was an assent to a person's testimony based upon a judgment of their character and a judgment of their ability. To "believe" something true was to trust the person who attested to its truth. If this trust were not a motive in the decision to assent, the assent fell short of being a species of belief.

Pearson so defined belief that the assent took no notice of the probability of its object. Such indifference was only rational if the believer could trust the knowledge which authorities claimed to have. Sprat and Locke denied that authorities could have knowledge which was absolutely trustworthy. No matter how certain they claimed to be about the things which they attested, there was always the possibility that authorities might be wrong. Both Sprat and Locke concluded that authorities could not be trusted beyond the probability of their testimony. This conclusion forced Sprat and Locke to redefine belief. They rejected Pearson's notion that the approval of credible authorities was the only legitimate reason for believing something true. Any reason was legitimate if it purported to show the truth of the object to be believed. The testimony of credible authorities was only one of many reasons for believing something true. Why exclude other kinds of evidence when awarding one's beliefs?

This chapter is a study of the meaning of belief in the thought of the clerical virtuosi. It seeks to measure the correlation between a cleric's definition of belief and his opinion of the certainty to which knowledge could attain. This chapter will begin by looking at the way two clerics, John Pearson and Isaac Barrow, defined the meaning of belief. Both were fellows of the Royal Society and successively presidents of Trinity College, Cambridge. Both defined the word, "belief," in respective expositions of the Apostles Creed.

Neither cleric was the author of the definition he expounded. Both borrowed their definitions from intellectual traditions which predated the Reformation and the rise of modern science. Pearson borrowed his definition from the logical tradition of medieval scholasticism. Barrow borrowed his definition from the rhetorical tradition of Renaissance humanism. Though both traditions assumed the possibility of absolute certainty, only Pearson's definition depended upon this assumption. It could only coexist with a theory of knowledge which admitted the possibility of absolute certainty. Barrow's definition of belief was far more flexible. It was able to coexist with a theory of knowledge which denied the possibility of absolute certainty.

John Pearson and the Meaning of Belief

John Pearson was an impoverished cleric when he published his *Exposition of the Creed* in 1659. Deprived of his prebend by the abolition of episcopacy in 1642, he forfeited his remaining livings by his decision to serve as chaplain to the Royalist general Lord Goring, during the First Civil War. The collapse of the Royalist forces in 1646, obliged him to retire to London where, subsisting on the proceeds from a small patrimony in Norfolk, he quietly pursued his scholarship until the Restoration.

Pearson successfully evaded his forced retirement only once. In 1650, the parishioners of St. Clement's, Eastcheap elected him lecturer. In offering him a lectureship, the parishioners of St. Clement's coopted an institution which the Puritans had once exploited as an instrument of dissent.[20] Though the post paid Pearson nothing, it did restore his right to preach—a right he had lost when the government sequestered his benefices. The possibility is remote that so staunch a Royalist could have regained a sounding-board for his opinions, had the parishioners of St. Clement's not intervened. Pearson made the most of the opportunity. The course of sermons he delivered during his term as lecturer formed the nucleus of the *Exposition*.

Upon the Restoration, Pearson was rewarded for his efforts on behalf of Church and King. Charles II made him Lady Margaret Professor of Divinity in 1661, and awarded him the masterships of Jesus and Trinity Colleges, Cambridge successively in 1661 and 1662. In 1673, he resigned his chair and mastership to succeed John Wilkins as bishop of Chester. Two years before the sixth printing of the *Exposition* in 1683, Gilbert Burnet judged him "the greatest Divine of the Age."[21]

Pearson continued to publish throughout the Restoration. His book on biblical chronology graced the shelves of Newton's library.[22] His defense of the authenticity of the epistles of St. Ignatius was a major contribution to the effort to square episcopacy with the primitive constitution of the early church.[23] Yet of all his works, Burnet considered his commentary upon the Creed "among the best that our Church has produced."[24]

When Pearson preached before St. Clement's in 1650-51, he was careful not to comment directly upon the Creed. Pearson later reported that he had merely expounded "such Texts of Scriptures as were on purpose selected in relation to the Creed."[25] To have preached openly upon the Creed in the early 1650's, would have implied criticism of the Creed's absence in the *Directory of Public Worship*, the Puritan successor to the *Book of Common Prayer*. A 1645 ordinance had set a fine of five to fifty pounds for preaching or writing against the *Directory*. The same ordinance had banned the *Book of Common Prayer* and any use of the *Directory* not in keeping with "the true intent and meaning thereof, in all exercises of the Publique Worship of God within this Realm."[26] Though the ban remained in effect until the Restoration, the early 1650's saw a relaxation in its enforcement as Cromwell sought to reconcile moderate episcopalians to the new regime. Until the Royalist uprisings of 1655, convinced Cromwell that reconciliation was impossible, private, even public, use of the prayer book spread. No one encouraged its spread more strongly than the "Laudians," Robert S. Bosher's term for the extreme episcopalians. Bosher places Pearson among their number.[27]

It is Bosher's contention that Cromwell's overtures to the moderates would have split the episcopal party had the Laudians not succeeded in making the prayer book a further stumbling block in the way of comprehension. The

Laudians, unlike the more moderate episcopalians, believed the new order in Church and State to be schismatic. They saw no room for compromise. When in 1653, the moderates petitioned the bishops for dispensation to use other forms of public worship than the prayer book, the Laudians made the argument that adherence to the *Book of Common Prayer* was of the same status as episcopal ordination. To the Laudians both were integral to the constitution of the Church of England. Accommodating either principle in the interests of comprehension was tantamount to apostasy.[28]

Though circumstantial evidence suggests that Pearson was acting partisanly in organizing his lectures around the Creed, there is little in the *Exposition* to support the notion that its author intended the Creed to be a symbol of dissent. In the preface which he dedicated to the parishioners of St. Clement's, he makes no mention of either the absence of the Creed in the *Directory of Public Worship* or the prayer book controversy of 1653. Rather than warn his parishioners about the dangers of making peace with a schismatical church, he bemoaned their lack of instruction in the fundamentals of the faith and predicted that their ignorance would lead them into heresy.

His concern for their orthodoxy may reflect the rate at which left-wing Puritans were adding numbers to their ranks. Both the early 1650's, when Pearson preached before St. Clement's, and 1659, the date when the *Exposition* was first published, were periods in which left-wing Puritans could proselytize at will without fear of persecution.

Pearson complained in his preface that the growth of heresy had reached such proportions that the very principles of Christianity were "now freely questioned as the most doubtful and controverted points."[29] Pearson blamed this state of affairs upon the arrogance of those who would "wrest the words of God himself unto their own damnation."[30] Such persons lacked respect for the authority of tradition, the judgments of Church Councils, and the opinions of the early Fathers. They interpreted the Holy Scriptures as if Christ had awaited their coming to found His true Church upon the earth. They even had the *hubris* to suppose that only now were the truths of Christianity being discovered to the world. Pearson reminded his readers that the truths of Christianity had shone forth from Scripture for the past seventeen hundred years. What these heretics were advancing was something new and "in Christianity," Pearson contended, "there can be no concerning truth which is not antient; and whatsoever is truly new is certainly false."[31]

Pearson presented the Creed as the most ancient and "probable guide" to the truths of Christianity.[32] Pearson did not argue that the Creed was apostolic. Gerhard John Vossius and Archbishop Ussher had exploded this pious legend some ten years earlier.[33] Pearson merely argued that no confession of the faith was more ancient or widely accepted than the Creed itself. Its antiquity and widespread acceptance "in all ages of the Church" made it the best compendium of the truths of Christianity.[34] He acknowledged the possibility that there were truths in Christianity that received no mention in the Creed, but he excused himself as an expositor of the Creed from making any determination of what these truths might be.

Pearson's defense of the Creed did not elevate the Creed to a position coequal with Scripture. The Scriptures remained for Pearson the ultimate source of the Christian faith. The Creed was only a guide to the contents of that faith. Pearson cautioned his readers that such a guide was only useful if it led them back "to the Scriptures, from whence it was at first deduced."[35]

Pearson set out to prove to the satisfaction of the heretics that the Creed was a true reflection of the Christian faith. He was prepared to dispense with tradition in arguing his case inasmuch as the Scriptures were the only authority his opponents would accept. Rather than place his Scripture proofs in the margins of his text, he endeavored to weave their language into the body of his text. He defended this practice by stating that "if those words which I have produced, prove not what I have intended, I desire not any to think that there is more in the places named to maintain it."[36]

Pearson opened the *Exposition* by attempting to explain what was meant in the Creed by the words, "I believe." He began by offering a general definition. Belief, he asserted, was "an assent to that which is credible as credible."[37] The definition is scholastic and at first glance uninformative. The definition tells one that belief is an assent, that the object of the assent is something credible, and that the credibility of the object is the reason for the assent, but it does not tell one what the conditions for credibility are nor what criteria may be used in measuring credibility. In the absence of such information, the term, "credible," a Latinism, is no more instructive of the nature of belief than its English equivalent, "believable." Substitute "believable" for "credible," and the definition states the obvious. Belief is merely an assent to something capable of being believed because it is worthy of belief.

The term, "assent," is similarly uninstructive. The definition declares that belief is an assent, but the term, "assent," cannot be the genus of "belief" because there are no assents which do not entail belief. "Assent" and "belief" are coextensive in their meaning, and thus practically synonymous. Both refer to that mental act whose aim or object is the truth. It is this commitment to the truth which distinguishes believing or assenting from such other mental acts as conjuring up images, searching one's memory, or directing one's thoughts.[38] To be told that "belief" is synonymous with "assent" may be of use to someone who knows the meaning of "assent" but not "belief," but to someone who is ignorant of either, it is the same as being told that "belief" is a "belief."

Given the triteness of the definition, it seems odd that the seventeenth century should have held it in high esteem. John Norris, one of Locke's clerical opponents, concluded his account of faith and reason with an apology for those places where he departed from "Authors of the better character...particularly Baronius and Dr. Pearson."[39] Norris need not have apologized for his definition of "belief;" it agreed with Dr. Pearson's. When the printer, Brabazon Aylmer, compiled a catechism in defense of the Trinity from the works of two influential divines, Bishops Tillotson and Stillingfleet, he did not use their definition of "belief," but quoted Pearson's verbatim.[40] Even Isaac Barrow, who rejected the definition, never argued that its triteness was the reason he rejected it. He rejected the definition because it was "restrained and artificial."[41]

This suggests that the seventeenth century understood the definition and found it to be informative. This fact is difficult to accept because it is difficult to believe that our assessment of the definition could be open to debate. Given the meaning of its terms, only one conclusion is possible. The definition is uninformative. This dilemma can be resolved if we assume that "belief" meant something different in the seventeenth century than it does today. To get at the meaning of "belief" in the seventeenth century, we must interpret the definition as if it were informative and deduce a meaning for "belief" that would satisfy the definition.

The definition would be informative if the term, "assent," were the genus of "belief" and the term, "credible," its difference. Belief would then become a species of assent. As a species of assent, it would continue to have the truth as its aim or object, but its commitment to the truth would cease to define its uniqueness as a mental act. The source of its uniqueness would become its credibility. Possession of this attribute would not only define its uniqueness as a species of assent, it would also differentiate belief from other species of assent like knowledge and opinion. Exactly how it would do this is not readily apparent. What does seem clear, however, is that belief would represent a distinct mode of cognition vis-a-vis knowledge and opinion. Knowledge, opinion, and belief would now differ in kind, not just degree.

Although this meaning of "belief" will satisfy the definition, what evidence do we have that this meaning was the one which Pearson actually intended? One source of evidence is the testimony of Pearson's contemporaries concerning the sense of his definition.

Though Barrow rejected the definition, his discussion of the definition supports this reading of the definition. Barrow ascribed the definition to "men of art"—a passing reference to its popularity among Protestant scholastics. Barrow reported that "men of art" divided all assents into several kinds "whereof they make Belief in one kind distinct from others." He further observed that they made belief "precisely denote that kind of assent, which is grounded meerly upon the authority (the dictate or testimony) of some person asserting, relating, or attestating to the truth of any matter propounded." According to Barrow, their intention in so doing was to distinguish belief from assents "grounded upon experience or appearance to sense; or upon rational inference." Though he criticized their action, he understood their motives and found them to be intelligible.[42]

John Norris defended an interpretation of belief very close to the one being attributed to Pearson. This can be seen in his warning against confusing belief with opinion. Norris thought the use of both terms interchangeably betrayed a lack of appreciation for the ground of their distinction as species of assent. Norris explained that "though there be a like uncertainty in both Assents, yet they differ extremely in their Formal Motives, one being grounded upon Reason, and the other upon Authority."[43] He argued that since "the Distinction of these Assents is not taken from the degree of Certainty wherein they agree, but from the Quality of the Motive wherein they differ," they should not be equated in ordinary usage.[44]

When we turn from Pearson's contemporaries to the *Exposition* itself, the text reveals that its author understood belief in the very way that Barrow depicted and Norris defended. Pearson argued that any definition of belief would be incomplete if it did no more than identify belief with assenting in general. Assenting he defined as "that Act or Habit of the Understanding, by which it receiveth, acknowledgeth and embraceth any thing as a Truth."[45] Though this would seem a perfectly adequate definition of belief, Pearson maintained that it was not, so long as it applied equally well to knowledge and opinion. As a species of assent, belief stood in need of further definition if it were not to be confused with other species of assent.

Pearson distinguished belief from other species of assent by focusing upon its reason, cause, or motive. He called the reason why we believe a thing, the "formal object" of belief. He termed the thing believed, the "material object" of belief.[46] Pearson argued that belief's formal object made the assent unique. While other assents could share belief's material object, none could share its formal object. Thus, belief's material object could be an object of knowledge or an object of opinion—depending upon the reason that prompted its assent. Pearson illustrated this point by offering the following example. He wrote:

> For he which sees an action done, knows it to be done, and therefore assents unto the Truth of the performance of it because he sees it: but another person to whom he relates it, may assent unto the performance of the same action, not because himself sees it, but because the other relates it; in which case *that which is Credible* is the object of Faith in one, of evident knowledge in the other.[47]

The phrase, "*that which is credible*," was Pearson's designation for the material object of belief. The next task which Pearson faced was to define for his readers the formal object of belief. He did this by comparing the formal object of belief with the formal objects of other assents. Pearson explained that objects which were apparent to either sense or reason could function on occasion as objects of belief, but if the motive for their assent rested upon their evidence to sense or reason, such objects were objects of knowledge, not objects of belief. Thus, "Snow is white" and "Fire is hot" were objects of knowledge because of their evidence to sense whereas "The whole of anything is greater than any one part of the whole" and "Every thing imaginable either is, or is not" were objects of knowledge because of their evidence to reason.

The next species of assent which Pearson introduced were those whose objects were the products of deductive reasoning. Although these objects were not "immediately apparent in themselves" like objects of knowledge, they still appeared "most certain and evidently true by an immediate and necessary connexion with something formerly known."[48] Such objects could not be said to be objects of belief if their "connexion" to objects of knowledge formed the motive for their assent. This motive, when it was present, made them objects of "science," not objects of belief.

The final class of assents which Pearson introduced were those whose objects appeared to most as true "by some external relations to other truths; but yet so, as the appearing truth still leaves a possibility of falsehood with it, and therefore doth but incline to an Assent."[49] Pearson grouped these assents under the category of "opinion." Their objects were "probable," not evident or "scientifical."

The assents he called "belief" were assents that fetched their motive not from the probability of their objects nor from their evidence to sense or reason, but from the credibility of the persons who attested to their truth. Objects of belief did not manifest their truth. Their truth was manifested by the testimony of others.

If this is the meaning of belief which Pearson actually intended, it affords few if any parallels with that of modern usage. No one today would argue when contrasting knowledge and belief that the act of knowledge excludes belief. Indeed, modern usage supports the notion that knowledge includes belief plus much more, namely, truth and good reasons.[50] Yet this notion would not follow if knowledge and belief actually differed in kind. As alternate modes of cognition, they would exclude each other.

Thus, to read Pearson's definition as if it were informative results in a definition of "belief" which makes no sense. His definition is unintelligible because the word he was defining no longer warrants his definition. To comprehend his definition, we must recover the warrant that allowed belief to become a species of assent. This task would be impossible if no parallel existed between the present meaning of "belief" and its meaning in the seventeenth century.[51]

One such parallel is Pearson's identification of belief with faith. For Pearson, to believe was to have faith and to have faith was to believe. Pearson never used "belief" to refer to assents other than "faith," as we do today. And yet there is a warrant in modern usage which would justify such a restriction. Modern usage allows that people as well as propositions can be objects of faith. The same holds true for belief. One can believe a person or believe a proposition. When directed toward a person it implies the element of trust, and when directed toward a proposition the element of trust is also present. When we have faith in a proposition, we believe it true because we trust the integrity and ability of the person who has attested to its truth. To "believe" a proposition upon any other motive is tantamount to "disbelief," for the element of trust is absent.

This can be illustrated by a stock example from the seventeenth century.[52] Suppose an astronomer told a rustic that the sun was larger than the earth. The very improbability of his testimony given the apparent size of the sun in relation to the earth might cause the rustic to doubt the astronomer and determine the issue for himself. If he reached the same conclusion as the astronomer, and answered him by saying, "I believe you," the astronomer might respond that he did not "believe" him, he only "agreed" with him, for if he had really believed him he would have accepted the astronomer's testimony upon his word alone.

The rustic had not done so because he allowed the improbability of what was being said to detract from the authority of the person who was saying it. The only criterion in matters of faith, and thus the only avenue for the exercise of doubt, was the credentials and intentions of the person giving the testimony. Once the rustic had satisfied himself concerning these, he should have trusted the astronomer despite his testimony.

It was reasonable for the rustic to trust the word of the astronomer if the astronomer had knowledge which was superior to the rustic's and if his knowledge could not be known by any other means than by his testimony alone. The first condition was met if there was reason to believe that the astronomer had certain knowledge about the thing he was attesting. The rustic's confidence that this was so could not be based upon the probability of his testimony or the assent that would result would lack the proper motive for belief. The rustic's confidence had to be based upon the word of the astronomer that he had such knowledge. The rustic could trust the astronomer if the astronomer had a history of telling the truth about things which the rustic could easily verify and no motive for deceiving him in the present instance. If his record of integrity and ability showed the astronomer to be credible, then the rustic was obliged to believe the astronomer upon the strength of his credibility. Though the rustic could not know that the object of his assent was absolutely certain, his belief in its necessity, based upon the word of the astronomer, was sufficient to overcome any doubts which might arise from the object itself.

The second condition was met if the rustic could not know, independently of the astronomer, whether the object of his assent was absolutely certain. If this condition were not met, it could still be argued that it was more prudent for the rustic to withhold his assent until he was absolutely certain of the truth of his assent. To assent to an object merely upon the word of the astronomer was to run the risk of error. Such a risk was only reasonable if the rustic lacked the means to verify the thing proposed for his assent. If the rustic lacked the learning to demonstrate its truth, he was obliged to trust the astronomer if he found him to be credible.

The rustic's obligation to believe the astronomer was increasingly questioned in the seventeenth century as doubts were raised about the certainty of the knowledge to which the astronomer could attain. If the astronomer could not have knowledge which was absolutely certain, why was the rustic obliged to trust him if he found his testimony to be improbable? Pearson never questioned whether authorities could have knowledge which was absolutely certain, though he acknowledged their fallibility when he compared their testimony with that of Revelation. God's testimony was certain because God by definition was absolutely credible. Human authorities were not so credible. This difference in credibility made for two kinds of faith: human and divine. Human faith was "an Assent unto any thing Credible merely upon the Testimony of man." Divine faith was "an Assent unto something as credible upon the testimony of God."[53] Pearson did not denigrate the value of human faith, for upon it were grounded not only the practical affairs of everyday life, but even the first principles of the sciences. Pearson wrote:

> And upon this kind of Faith we proceed in the ordinary affairs of our life; according to the opinion we have of the ability and fidelity of him which relates or asserts any thing we believe or disbelieve. By this a friend assureth himself of the affection of his friend: by this the Son acknowledgeth his Father, and upon this is his obedience wrought. By virtue of this *Humane Faith* it is that we doubt not at all of those things which we never saw, by reason of their distance from us, either by time or place. Who doubts whether theire be such a Countrey as *Italy*, or such a City as *Constantinople*, though he never pass'd any of our four Seas? Who questions now whether there were such a Man as *Alexander* in the East, or *Caesar* in the West? and yet the latest of these hath been beyond the possibility of knowledge of man these sixteen hundred years. There is no Science taught without original belief, there are no Letters learnt without preceding faith. There is no Justice executed, no commerce maintained, no business prosecuted without this; all secular affairs are transacted, all great achievements are attempted, all hopes,

> desires and inclinations are preserved by this Humane Faith grounded upon the Testimony of man.[54]

In contrast to human faith, divine faith was never subject to error or betrayal. Divine faith was certain whether its contents were apprehended by an "immediate" revelation or whether its contents were apprehended by a "mediate" revelation. Pearson distinguished the two by saying that in an immediate revelation "God delivereth himself to man by himself without the intervention of man," whereas in a mediate revelation God conveyed himself "unto man by man."[55] The person who was granted an immediate revelation did not need to be persuaded of the source of that revelation. The source of the revelation was self-evident to that person. Only the person who heard God's word from the mouth of another person stood in need of further proof of the source of that revelation. This proof was provided by the testimony of miracles. Miracles attested that the person who performed them was a messenger from God and had the approval of God Almighty for the things they were revealing.

Pearson also pointed out that the things which God revealed could only be apprehended through His testimony alone. Pearson wrote:

> To which end it will be necessary to give notice that the *Testimony* of God is not given to truths before questioned or debated of by man, and then resolved and confirm'd by interposing the authority of God: but he is then said to witness what he doth propound, and his *testimony* is given by way of *Revelation*, which is nothing else but the delivery or speech of God unto his Creatures.[56]

Thus the truths of Revelation fulfilled all the conditions for rational belief. They could only be known through God's testimony alone and they could claim a certainty which was superior to human knowledge. Since the Creed was a summary of the truths of Revelation, Pearson assumed that its articles were absolutely certain. This assumption undergirded his full definition of what was meant in the Creed by the words, "I believe." Pearson wrote:

> To believe therefore as the word stands in the Front of the CREED, and not only so, but is diffused through every Article and Proposition of it, is to assent to the whole and every part of it, as to a certain and infallible truth revealed by God, (who by reason of his infinite knowledge cannot be deceived, and by reason of his transcendent holiness cannot deceive) and delivered unto us in the Writings of the blessed Apostles and Prophets immediately inspired, moved and acted by God, out

> of whose Writings this brief summ of necessary Points of Faith was first collected.[57]

Was it rational to give the Creed the same degree of faith as the truths of Revelation? Pearson thought it was if the Creed truly mirrored what was contained in Revelation. Tradition had attested that the Creed was such a mirror, but the only way to prove it was to compare its content against the content of Holy Scripture. Pearson proposed to do so by deducing its articles from the pages of Holy Scripture. Was such a proof a demonstration given the uncertainties involved in interpreting Holy Scripture? Pearson never asked this question. It was foreign to his outlook as a Protestant scholastic. Only Barrow asked this question and he answered it in the negative. His answer helped determine the way he conceptualized belief.

Isaac Barrow and the Meaning of Belief

On Easter Eve 1669, Isaac Barrow wrote John Collins and apologized for his failure to send several new "geometrical theorems of a general importance." He explained that he had intended to send them, but had been "hindered" from doing so by the demands "imposed upon me by the college." In addition to his "pupils and other ordinary employments," he had been required to compose a set of "Theological Discourses (as our Statutes order) upon the chief points of Catechism (the Creed, Decalogue, Lord's Prayer, Sacraments, &c.)." These he was composing "out of term" and he gave Collins to understand that they had so taken up his thoughts that he had had no time for any other matters.[58]

It is commonly supposed that Barrow composed these "discourses" in order to qualify for the office of college preacher. Since Barrow was appointed to this office in 1671, he probably delivered these discourses in the chapel of Trinity College, Cambridge sometime shortly before this date. If so, it is very likely that John Pearson, then master of Trinity, was a member of the audience that heard Barrow preach upon the first article of the Creed.

When Barrow composed the "discourses" that he preached in Trinity Chapel, he was on the verge of resigning his Lucasian chair of mathematics. Not only did the resignation mark a shift in his interests from mathematics to theology, it also freed him to campaign for posts like that of college preacher. The statutes governing his chair had hitherto precluded his accepting such preferment. The "discourses" he composed were the first shot in a campaign that culminated in his appointment as president of his college.

Though the campaign proved successful, it entailed financial hardship. With only the proceeds of his Trinity fellowship to satisfy his needs and those of his relations, whom Abraham Hill reports were in a "necessitous condition," Barrow had to depend upon the charity of his friends to make up the difference. His uncle, Isaac Barrow, bishop of St. Asaph, gave him a small sinecure in Wales and his friend, Seth Ward, bishop of Salisbury, welcomed him into his household and gave him a prebend in his cathedral.[59]

Walter Pope, Ward's half-brother, attested to Barrow's "indefatigable Study" in this period. Pope reported that at the time Barrow "applied himself wholly to Divinity, having given divorce to Mathematics, and Poetry, and the rest of the *Belles lettres*, wherein he was profoundly versd, making it his chief, if not only business, to write in defense of the Church of *England*, and compose Sermons, whereof he had a great store, and, I need not say, very good."[60]

Barrow's move toward theology received an additional spur from the reception accorded his earlier work in mathematics. At the start of his campaign for university preferment, Barrow published his lectures on optics and geometry, the fruit of his tenure as first Lucasian professor of mathematics. Neither set of lectures received the acclaim that Barrow hoped for. Abraham Hill reported that Barrow knew of only two people who had read the lectures on geometry all the way through. Hill speculated that the "little relish that such things met with did help to loosen him from these speculations, and the more engage his inclination to the study of Morality and Divinity."[61] Barrow himself acknowledged the extent of his disappointment in the Latin verses which accompanied his gift of the *Lectiones Geometricae* to John Tillotson. Barrow wrote,

> While you, my dear friend, deliver the mysteries of sacred truth to the people in powerful eloquence; and shut the mouths of petulant sophists, and carry on successful controversy for the Law of God; I, as you perceive, am so unhappy as to be fixed to the books which you see here; and so waste my time and power.[62]

Barrow's literary efforts, his reputation as a scholar, and his cultivation of influential patrons finally bore fruit in 1673, when the King appointed him to the mastership of Trinity. Hill assured his readers that "it was the King's own act" though he conceded that Barrow's "desert made those of the greatest power forward to contribute to it, particularly Gilbert, Lord Archbishop of Canterbury and the Duke of Buckingham, then Chancellor of Cambridge, and formerly a member of Trinity College."[63]

Barrow's erudition was both his strength and his weakness. His patron, Charles II, once made the sardonic observation that Barrow "was the most unfair preacher he knew, for he never left anything for any one else to say on subjects which he handled."[64] Walter Pope made the same complaint when he observed that Barrow probably thought that "he had not said enouf, if he omitted any thing that belonged to the subject of his Discourse, so that his Sermons seemd rather complete Treatises, than Orations, designd to be spoke in an hour."[65] Though it is a misnomer to call his discourses upon the Creed, "brief," as Brabazon Aylmer did when he published them posthumously in 1697, their very copiousness is helpful in providing clues to Barrow's understanding of the meaning of belief.

Barrow began his explication of the first article of the Creed by considering the notion of belief which he intended to reject. Barrow observed that there was a "restrained and artificial acception" of the word that was "peculiar to men of art, and invented by the school," which limited its meaning to that kind of assent "which is grounded meerly upon authority (the dictate or testimony) of some person asserting, relating, or attesting to the truth of any matter propounded."[66] Barrow reported that the "person asserting, relating, or attesting" was termed "credible" if their authority was "apt in a well-disposed understanding to beget such an assent to the truth of what is deposed."[67] Such a person had authority if there was reason to believe that they had attested to the thing which was credited to them, that they had the credentials and qualifications to know what they attested, and that they had no motive to deceive in attesting to its truth. Barrow explained that the strength of one's belief varied in direct proportion to one's confidence that these three conditions had been met. Barrow wrote:

> ...according to the degrees of our assurance either concerning the fact, that the Author doth indeed assert the matter; or concerning the person's qualifications (rendring his authority credible) are the degrees of our Belief proportioned; it is more strong and intense, or weak and remiss; we are confident or doubtfull concerning the matter; if we plainly can perceive by our sense, or have great rational inducements to think, that such an assertion proceeds from such an Author; and then by like evidence of experience or reason are moved to think him not liable

> himself to be deceived, nor disposed to deceive us, then we become strongly persuaded; believe firmly, in proportion to the validity of the said grounds.[68]

Barrow contended that this "restrained and artificial" definition of belief compared unfavorably with the "more general and popular" one which he was recommending. Barrow asserted that belief generally signified "being well persuaded, or yielding a strong assent unto the truth of any proposition." Barrow asserted further that common usage did not distinguish between the means of persuasion that were properly said to "beget" this assent, but rather assumed that the assent could be produced by any means whatsoever.[69] Thus, a person could be said to believe "what our sense represents, what good reason inferrs, what credible authority confirms unto us." All were valid reasons for believing, not just the last.[70]

Thus, Barrow was convinced that the "more general and vulgar notion" of the meaning of belief was the proper interpretation of the first two words of the Creed. Barrow contended that when we say we believe in the articles of the Creed:

> ...we profess to be persuaded in our minds, concerning the truth of the propositions annexed, not implying our persuasion to be grounded upon onely one kind of reason, that drawn from authority, but rather involving all reasons proper and effectual for the persuasion of all points jointly, or of each singly taken.[71]

Barrow's reasons for selecting this "more general and vulgar notion" were as follows. It seemed improbable to Barrow that the "ancient teachers of our Religion" would have used the word, "belief," in any other sense than its "most common and familiar acceptation." They themselves were men "not seen in subtility of speculation, nor versed in niceties of speech (used by men of art and study.)"[72] And since their mission was "to instruct the generality of men," they could hardly hope to persuade them if they did not speak a language which they could understand. Barrow also pointed out that it was anachronistic to assume that the first century understood this scholastic definition since it was "devised too after their times."[73]

Barrow went on to argue that not only was it improbable that the "ancient teachers of our Religion" would have used the word, "belief," in any other sense than the one he was recommending, but the way it was actually used by them in Scripture showed that they meant belief to "signify indifferently all kinds of persuasion, having no regard to the particular ground thereof."[74] Thus, Barrow wrote:

> *Thomas* would not believe that our Saviour was risen, except he discerned visible marks, distinguishing his person from others: he did so, and then believes: whereupon our Saviour saith: *Thou believest, because thou hast seen; blessed are they that have not seen and yet have believed*: We see that faith may be grounded upon sense. And, *If I do not the works of my father* (saith our Saviour) *believe me not; but if I do, though you believe not me, believe the works.* Our Saviour requires them not to rely upon his bare testimony concerning himself, but to consider rationally the quality of his works; and upon that to ground their faith: which kind of persuasion seems grounded rather upon principles of reason, than any authority. The Devils (*St. James* tells us) do believe *there is one God*: how so? Because they know it by experience, rather than upon any relation or testimony given them.[75]

Turning to the Creed itself, Barrow argued that belief in its first and main article, namely, the existence of God, could not be grounded upon authority alone, for "humane authority cannot alone suffice to prove so great a point; and divine authority doth pre-suppose it."[76] And even though the truth of the articles of the Creed rested upon divine authority, God's authority alone could not prove that God was veracious or that he had attested to the articles which claimed to have His credit.

For all these reasons, Barrow concluded that the words, "I believe," which prefaced the articles of the Creed, should be interpreted to imply a "mind fully convinced and persuaded of the truth of the propositions hereafter expressed (or implied;) not excluding any objects there contained under any formality (either of being apparent to sense or demonstrable by reason, or credible by any sort of testimony) nor abstracting from any kind of reasons persuasive of their truth."[77] Thus, unlike his master, John Pearson, Barrow refused to discriminate between the means of persuasion that were appropriate in belief. Assents whose motive was their evidence to sense or reason were properly termed beliefs. So were assents whose motive was their immediate and necessary connection to truths of fact or truths of reason. So too were assents that were accorded to their objects for reasons that had to do with their probability or likelihood. Indeed, Barrow called "belief" every species of assent which his master had formerly distinguished.

In rejecting an interpretation of the Creed which defined belief as a particular kind of assent, was Barrow aware that he was rejecting an interpretation favored by the master of his college? Though Barrow is silent, there is evidence to suggest that he was familiar with Pearson's work. Earlier in his *Exposition*, Barrow cited a "late excellent exposition of the Creed" to support his view that the addition of the proposition, "in," did not affect the sense of "believe" in the first article of the Creed, "I believe in God."[78] Barrow's discussion of this question closely followed that of Pearson.

Barrow was not the first to reject the scholastic definition of belief favored by his master. In some respects Barrow's rejection of the scholastic definition followed a tradition in English thought that stretched back at least as far as Richard Hooker (1534?-1600). Hooker too asserted that "whatsoever we are persuaded of, the same we are generally said to believe," and he attacked those who wished to limit the means of persuasion in matters of belief to the word of authorities alone. Appealing to Scripture and common usage, Hooker contended that "we are sayd not onely to beleeve the thinges which we knowe by anothers relation, but even whatsoever we are certainly perswaded of, whether it be by reason, or by sense."[79]

Thomas Jackson (1579?-1640) held similar convictions about the meaning of belief. He too was not willing to limit its meaning to "that assent or perswasion, which relies upon authorities." Jackson contended in a commentary upon the Creed, first published in 1613, and reissued again in 1673, that the credibility of authorities was not the only motive in matters of belief. He cited the fact that "that which we evidently know, may oft-times be the cause, why we believe some other matters, that have affinity with it." Thus, "he who seeth it very light in the morning when he first openeth his eyes, may probably beleeve the Sunne is up, because he evidently seeth the ayre to be light." This belief rested upon grounds that were independent of authority. It was Jackson's opinion that "even those assents or perswasions, which seeme most to relie upon authorities" could be "strengthened" by "motives and inducements" from sense and reason.[80]

Nevertheless, Jackson was not willing to extend the meaning of belief to comprehend those "assents or perswasions" which were "evidently certain."[81] He explained that it was one thing to believe the sun was up because the air in the room was light, "but no man (if you should aske him the same question) would say, that he beleeved the Sunne was up, when either the heate thereof doth scortch his face, or the beames dazell his eyes; for now he knoweth this truth directly and evidently in it selfe."[82] Jackson concluded that "evident certaintie, either of sense or understanding, must be excluded from the assent, which is properly called beliefe" inasmuch as "evident certaintie doth drowne beliefe."[83]

Barrow's willingness to group assents or persuasions which were evidently certain under the denomination of belief suggests that Barrow held convictions about the relation between knowledge and belief that differed markedly from Jackson's. Jackson assumed that knowledge and belief denoted two cognitive attitudes or species of assent which were mutually exclusive. The person who was said to know could not be said to believe, and the person who was said to believe could not be said to know. To say that a person believed implied that the thing which they believed was capable of doubt whereas to say that a person knew implied that the thing which they knew was not capable of doubt because it was evidently certain. Though some things could be certain without being evidently true, (e.g., the mysteries of Revelation), nothing could be evidently true without being absolutely certain.

The seventeenth century witnessed an intellectual movement that questioned the automatic juxtaposition of the evident with the certain. This movement acknowledged that an object could be evident and still be false. The possibility was not removed because an object was so evident that it was impossible to deny. There was always the possibility that another object could exist which would contradict its truth. The fact that an object was evident was a sign of its truth, but it was not a necessary sign. Even objects which were evident were capable of doubt.

One of the effects of this movement was to change the way that the relation between knowledge and belief was conceived. William Chillingworth (1602-1644), writing a quarter century after Jackson, conceived of the relation as one of species and genus. Chillingworth conceded that properly speaking faith was not knowledge, "no more than three is four," but he argued that faith was "eminently contained" in knowledge as three is contained in four. It was Chillingworth's opinion that "he that knows, believes and something more; but he that believes many times does not; nay, if he doth barely and meerly believe, he doth never know."[84] Thus, belief and knowledge were not mutually exclusive.

Barrow's friend, John Tillotson, who was editor of his posthumous works, argued that faith "is not opposed to error, and knowledge, and opinion: but comprehends all these under it." Tillotson wrote:

> For if a man be persuaded of that which is false, he 'believes a lie,' as the Scripture expresseth it; a man may be certainly persuaded of a thing, that is, firmly believe it, which is *knowledge*; a man may be probably persuaded of a thing, that is, believe it with some diffidence and uncertainty, and that is *opinion*.[85]

Tillotson was able to conceive of knowledge as a species of belief because he no longer conceived of knowledge as absolutely certain. Absolute certainty, for Tillotson, was "an incommunicable attribute" belonging only to God. The mortal who claimed it "hath not the modesty of a creature, but doth by a sacrilegious ambition attempt the throne of God, and equal himself to the Most High."[86] Man could be certain, but he could not be absolutely certain. John Wilkins, Tillotson's father-in-law, defined the difference in the following way.

> By *Infallible Certainty*, I do not mean *absolute* Infallibility, because that is an *Incommunicable Attribute*. And it were no less than a blasphemous arrogance, for any man to pretend to such a perfect unerring judgment on which the Divine power it self could not impose. But I mean a *Conditional Infallibility*, that which supposes our faculties to be true, and that we do not neglect the exerting of them. And upon such a supposition there is a necessity that some things must be so as we apprehend them, and that they cannot possibly be otherwise.[87]

The conditionally infallible might compel an assent, but it could not banish the possibility of falsehood. Since the assent carried with it the possibility of falsehood, the assent was technically a species of belief, not a species of knowledge. Though Wilkins and Tillotson followed common usage in calling the conditionally infallible, knowledge, the assent was technically belief under criteria set forth by Aristotle. According to Aristotle, no knowledge was "scientific" unless the person assenting was certain that the reason why he assented was the reason why the thing, the object of his assent, was true. Perhaps even more important, no knowledge was really knowledge unless the person assenting was certain that the thing was so and could not be otherwise. If a person's "certainty" about an object was dependent upon their supposing that their senses and their reason were reliable sources of information, then that person could not be said to possess the absolute certainty which Aristotle required of "science."

Both Wilkins and Tillotson agreed that no assent could be so evident that it escaped the realm of the conditionally infallible, for all assents depended upon the assumption that what appeared to be true to either the senses or the understanding was actually the case. Both denied that any criterion existed that could tell a person when they were absolutely certain concerning the truth of any object.

Though Barrow nowhere in his *Exposition of the Creed* or his sermons upon the Creed discusses the level of certainty to which a person could attain, there is reason to believe that Barrow would have joined Tillotson and Wilkins in limiting mankind to the conditionally infallible. We have already seen that Barrow did not exclude from the title of belief assents which were "apparent to sense or demonstrable by reason."[88] The latter included the kind of "scientific" knowledge which Aristotle had assumed was infallibly certain. Barrow rejected the notion that some objects of assent were so evident that they necessitated one's assent. It was Barrow's opinion that belief was not "an effect of blind necessity," nor could a person "be compelled to believe any thing."[89] Belief was the "result of judgment and choice, grounded upon reason of some kind, after deliberation and debate concerning the matter."[90] This was as true for assents whose objects were self-evident as it was true for assents whose objects were merely probable. Thomas Jackson had assumed that some things were so evidently certain that "scarcely can any distemper of body, or mind, phrenzy excepted, work any distrust, whether the whole be no bigger than its part, *whether all right angles are equal, or whether ademption of equal portions from things equal, leave not such equality betwixt them as it found.*"[91] Barrow thought no thing however evident was incapable of doubt. In the course of arguing that the truths of the Christian religion carried with them sufficient evidence to justify an assent, Barrow argued that even if God "stamped on his truth that glaring evidence, which might dazzle our minds," there would still have been men who would have chosen to "outface and outbrave it, however clear and evident."[92] Barrow wrote of such men,

> They may pretend, if they had more light, they would be persuaded; like those who said, *Let him now come down from the cross, and we will believe*; but it would not in effect prove so, for they would yet be devising shifts, and forging exceptions; or, however, they would oppose an impudent face and an obstinate will against the truth.[93]

Barrow's stand on this issue is not entirely consistent. In his earlier lectures on the usefulness of mathematical learning, he used the traditional dichotomy between knowledge and belief to recommend the study of mathematics. Barrow contended that only mathematics could claim to be a "science," for mathematicians:

> ...affirm nothing among their Arguments or Assertions which is not most manifestly known and examined with the utmost Rigour, rejecting all probable Conjectures and little Witticisms. They submit nothing to Authority, indulge no Affection, detest the Subterfuges of Words, and declare their Sentiments, as in a Court of Judicature, *without Passion, without Apology*; knowing that their Reasons, as *Seneca* testifies of them, are not brought *to persuade, but to compel.* And by this means they preserve their Sciences pure and free from the least Stain either of Error or Doubt. In others, which are called Sciences, Things doubtful are promiscuously mingled with Things certain, Things taken upon Presumption with Things found by Experiment, Things obscure with such as are manifest, and true Things with false. Their Professors propose all Things to themselves to be examined, explained, and decided, according to what they know and what they do not know: And when they cannot bring certain Reasons, they produce any Kind of Conjecture, disguised under beautiful Colours, oftener divining than demonstrating. In find, they plead with all their Eloquence, as for their Properties; and what they can't prove, they endeavour to persuade.[94]

In the quotation above, Barrow opposes persuasion, error, and doubt to compulsion, certainty, and proof. In a paean to the Lord, as the great Geometer, which he inscribed on the flyleaf of his copy of his edition of *Apollonius*, Barrow compared geometry with other subjects:

> In other subjects our Intellect can do almost nothing; like the phantasy of brute creatures, it seems only to present some doubtful and dreamy notions, and thus there are as many opinions as there are men. But in this subject all agree: in this the human mind sees that it can do something; something even great and wonderful; wonderful indeed, for the faculty which is in other things so helpless is here effective and successful.[95]

Barrow praises geometry because it is a subject in which all agree. Whether this agreement is the result of the necessity of its subject matter is only hinted at, but elsewhere Barrow appears to be suggesting this. Referring to the nature of faith as a voluntary act, Barrow wrote,

> Indeed to prevent this exception, that faith is a forced act, and therefore not moral; or to render it more voluntary and worthy, God hath not done all that he might have done to convince men, or to wring belief from them: he hath not stamped on his truth that glaring evidence, which might dazzle our minds; he doth not propose it armed with irresistible cogency; he hath not made the objects of faith conspicuous to sense, nor the propositions thereof demonstrable by reason, like theorems of geometry:. . .[96]

If the theorems of geometry have irresistible cogency, does this mean that they necessitate assents, and if they necessitate assents does this mean that they are infallibly certain and beyond the least stain of error or doubt? Barrow does not say, but he implies it. On the other hand, is the relation between persuasion and compulsion the same as that which Chillingworth sketched out for knowledge and belief? Is compulsion a kind of persuasion? It would seem to be, since Barrow is adamant in not excluding any means of persuasion from the assent called belief—and this includes assents which are evident to sense or demonstrable by reason. Perhaps all that Barrow is arguing is that even these assents require an act of judgment. This argument can be found in Aristotle. But Barrow goes further. He says they require deliberation and choice, and Aristotle says that no one deliberates about things which do not seem to admit of issuing in two ways. This would indicate that perhaps nothing is so evident that it escapes the possibility of issuing in another fashion than the way we perceive it.

Scholasticism and Humanism In the Thought of Pearson and Barrow

In conclusion, another way of looking at the differences between Pearson and Barrow is to set them in their historical contexts. Pearson's definition of belief is scholastic and its derivation is medieval, whereas Barrow's definition of belief is humanistic and it derives from the Renaissance. The labels, "scholastic" and "humanist," when applied to Pearson and Barrow convey more of the thrust of their respective theories of belief than the labels, "Laudian," "Latitudinarian," and "Puritan," which carry a sociological import that is curiously inappropriate in this context.

Scholasticism and humanism continue to frustrate attempts to define them precisely. Both movements predate the Reformation and thus there may be something misleading in calling two late seventeenth-century Protestant divines "scholastic" and "humanist." Brian G. Armstrong concedes that scholasticism "defies precise definition." He contends that it is a matter of "more-or-less identifiable tendencies," an "attitude to life" rather than a "list of beliefs."[97] Stephen Ozment agrees. He contends that "humanism and scholasticism not only defy simple, solitary definitions," but also a "prevalent scholarly tendency to depict them as mortal enemies."[98]

What then are these tendencies or distinguishing characteristics? Ozment sees scholasticism's primary concern with the definition and defense of doctrine. In its concern with the truth of doctrine, scholasticism is less concerned with the ethical implications implicit in a doctrine. What is important to scholasticism is the unity of doctrine and the continuity of its historical interpretation. Truth for the scholastic remains the same in all ages, and if one wants to know the truth, consult what has been affirmed in all ages concerning the truth. It is important to recognize here that a concern with tradition—something which all scholastics exhibited in their works—did not translate into a concern with history, if history is defined as an account of change over time. Truth did not change but remained the same. Scholastics were concerned to minimize evidence of change in regard to doctrine. Change meant declension from the truth, departure from the unity of tradition. The principles of faith remained unchanged though every age might discover new riches in these principles through their logical elaboration. In looking back over tradition, the scholastic doctor tried to reconcile divergent opinions to demonstrate that tradition indeed had spoken with one voice concerning the truth.

In its overriding concern with the definition and defense of doctrine, scholasticism displayed an attitude of mind that seemed more interested in proof than persuasion. Scholasticism believed that if it only stated the truth, everyone would agree. At least humanism recognized that truth had a need for the arts of persuasion. Most truths were not so evident that they necessitated assents. Most were subject to opinion. Aristotle, unfortunately, never developed an adequate logic for measuring the certainty of truths in matters of opinion. The only criterion he could offer was their capacity to persuade. The more approval a truth received the greater its probability and the greater its probability the more likely it was true. Was the approval a truth received an adequate criterion of its truth and certainty? Humanists often argued as if it were.

Their arguments also confused questions of truth with the questions of prudence. They often appeared to argue that the conditions which made it reasonable to act upon a thing were the very same conditions that made it reasonable to believe it true. Yet belief in Christianity was not a matter of utility. Their confusion can be traced to their conception of belief. They made belief in Christianity synonymous with persuasion. When they identified belief with the act of persuasion they overlooked the fact that the end of all persuasion was action, not the truth. Truth in matters of persuasion was only a means of persuasion, for the goal of all persuasion was right action, not right belief. Thus the humanists were concerned to defend the truth of Christianity to the extent that its truth prompted obedience to its precepts.

Of the tendencies associated with the scholastic outlook, Armstrong lists the "sentiment that the scriptural record contains an unified, rationally comprehensible account" of the Christian faith that is capable of being formed into a definitive statement of orthodoxy. Armstrong argues that this sentiment arose from a particular theological approach which asserted "religious truth on the basis of deductive ratiocination from given assumptions or principles, thus producing a logically coherent and defensible system of belief." Armstrong adds that this generally "takes the form of syllogistic reasoning," an orientation he attributes to the scholastic commitment to Aristotelian philosophy.[99]

Although Armstrong is correct, it is misleading to make scholasticism the logical result of an adherence to Aristotelian philosophy. In most fields of human endeavor in the seventeenth century there was no alternative to Aristotle. E.J. Dijksterhuis writes:

> The influence that Aristotle has exercised on scientific thought is not entirely due to the study of his works on metaphysics, physics, and astronomy. In the collection of his writings on logic, which is usually called the *Organon*, there are general epistemological reflections on the requirements a demonstrative science should fulfill, and in fact during the renaissance of scientific thought in the sixteenth and seventeenth centuries these ideas, instead of being confuted and rejected together with his ideas on nature seem to have served precisely as the guiding principle for the foundation of mechanics and that part of physics which is based on it.[100]

Humanists as well as scholastics accepted the validity of Aristotelian logic and thought all sciences should conform to the pattern of presentation set forth in the *Posterior Analytics*. It is not an adherence to the Aristotelian inheritance which differentiates scholastics from humanists, but the weight they placed upon different aspects of this tradition.

This matter calls for illustration. As a scholastic, Pearson believed that theology constituted a divine science. The theologian abstracted his first principles from Scripture and by means of syllogistic reasoning explored their implication for life and doctrine. The first principles of the theologian were no different than the principles of any other science except for the fact that these principles were inevident. Though they stood in contrast to the self-evident first principles of *scientia*, both had to be taken on faith. To demand proof of the first principles of science would lead to an infinite regress. To demand proof of the first principles of theology was to ask what could not be given. Both sets of principles demanded faith. (Indeed, in some respects theology had an advantage over the human sciences since it received its first principles from God who could not lie or be mistaken about the things He was attesting.)

Where a humanist like Barrow differed from a scholastic like Pearson was over the process of abstracting those first principles from Scripture. As a scholastic, Pearson had a greater estimation of the uses of tradition in the interpretation of Scripture than Barrow. This arose from his belief in the unity and immutability of truth, mentioned earlier. As a good Protestant, Pearson believed that Scripture and Scripture alone contained the first principles of the Christian faith. Tradition attested what Scripture affirmed; it was not coequal with Scripture as an authority. Nevertheless, to interpret Scripture apart from tradition was to do so at one's peril.[101]

Pearson believed that without tradition, he had no way of telling whether the first principles he abstracted were really contained in Scripture. Unless his principles corresponded with the ones that tradition had always found in Scripture, he had only his persuasion to prove that they were true. If his own contradicted those affirmed by tradition, he had to resolve the contradiction, and if his own turned out to be different, he had to show how they were corollaries of what had always been affirmed.

Barrow had less respect for tradition and it is important that we understand why. Neither Barrow nor Pearson would have thought that nothing more was necessary to understand the Bible than a "vulgar Translation, and a large Concordance."[102] The abstraction of first principles from the biblical text remained as problematic for the humanist as it was for the scholastic—perhaps even more so since humanists were unable to avail themselves so easily of tradition. Tradition had commented upon a text which humanist biblical scholarship had shown to be defective and full of textual errors. What authority were the affirmations of tradition if they were based on a mistranslation? Humanists brought to their interpretation of the Bible an increasing appreciation of historical change. By stressing the unity and immutability of truth, scholastics like Pearson had approached Scripture a-historically. William Payne, another clerical virtuoso, succinctly expressed the humanist attitude. Payne wrote:

> ...Religion that was taught by God, and deliver'd from Heaven is contain'd and transmitted down to us in such Books and Records as do necessarily require Learning to understand them, not only the skill in Languages as a key to unlock and decypher the meaning of these Holy Books, and the peculiar Idiotisms and Proprieties of those Languages, without which we have only the sound and not the sense of them, but the knowledge of Lives and of Places, and the Customs and Manners of those People to whom many things in Holy Scripture do immediately relate, and their true meaning cannot be understood without them; one of the great Causes of mis-interpreting, and mistaking the sense of Holy Scripture, is the taking it as a Book writ in our Times, and not many Ages ago,...when some of the most considerable things in it, not only the Prophecies of the Messiah and the time of his coming, but the nature of the two Sacraments nay of that of our Saviour's Sacrifice and satisfaction, cannot well be understood without knowing the Jewish and the Heathen Notion about Sacrifices, and yet 'tis that upon which the whole Gospel economy does mightily depend.[103]

To appeal to tradition in defense of an interpretation of Scripture was to beg the question in cases where philological and historical criticism revealed that the Scripture cited was a mistranslation or one whose sense tradition had taken anachronistically. Whereas tradition had been the rule of Scripture, increasingly Scripture—subjected to philological and historical analysis—became the rule of tradition. We have seen a good example of this in Barrow's rejection of Pearson's scholastic definition of faith as unscriptural. Without the help of tradition, the enterprise of abstracting the first principles of Christianity from Scripture became a more difficult and uncertain venture. Humanism reasoned that if scholasticism could never obtain for the theology the status of a science, why use the scholastic way of science in the elaboration of theology? If the first principles of theology could be no more than probable, then it seemed foolish to spin forth systems which could hope to be no more than probable. It would be far better and certainly more prudent to remain close to Scripture, assenting to those principles which from Scripture one had most confidence one had understood correctly. The old boast that theology was as certain as any of the other sciences appeared increasingly empty in the light of these textual problems. Theology was only as certain as one was certain that what one took to be the Word of God was indeed the Word of God, and that what one deduced from Scripture was indeed rightly deduced.

In this attitude, we can discern a powerful impulse toward moralism in theology. If, as Payne intimates, the chief doctrines of Christianity could not be understood without an understanding of the languages and culture out of which such doctrines came, then the only doctrines that appeared to survive this historicism with their meaning intact were the moral codes. Payne states that "the great and necessary lines of Morality are so plain" not only in Scripture, "but even on the hearts of all Men that they need not much Comment."[104] The moral precepts of Christianity, unlike its metaphysical doctrines, were unambiguous and clear to all. Humanists yielded to the temptation to identify these precepts as most basic to Christianity. One needed learning to correctly understand Christ's nature, his purpose for coming into the world, and the nature of his atonement; one needed none to understand the golden rule and the Sermon on the Mount. Living a Christian life, rather than understanding what it asserted about things metaphysical and above reason, was the essence of Christianity. If this were not the case, what salvation could there be for the great unwashed who were unable to properly conceive the abstruse doctrines of Christianity?

NOTES

1 Thomas Sprat, *History of the Royal-Society of London, For the Improving of Natural Knowledge*, ed. Jackson I. Cope and Harold Whitmore Jones (London: Printed by T.R. for J. Martyn at the Bell without Temple-bar, and J. Allestry at the Rose and Crown in Duck-Lane, Printers to the Royal Society, 1667; reprint ed., St. Louis: Washington University Studies, 1958), pp. 47-48.

2 Ibid., p. 49.

3 Ibid., p. 50.

4 Ibid., p. 49.

5 Ibid., pp. 5-6.

6 Ibid., p. 30.

7 Ibid., p. 21.

8 Ibid., pp. 29-30.

9 Ibid., pp. 99-100.

10 Ibid., pp. 99-100.

11 Ibid., pp. 48-49.

12 Ibid., pp. 47-48.

13 John Locke, "Of Study," in *The Educational Writings of John Locke*, ed. James L. Axtell (Cambridge: Cambridge University Press, 1968), p. 408. On p. 405, Axtell informs the reader that Locke composed this essay from the 26th of March to the early part of May 1677, while traveling in France.

14 Ibid.

15 John Locke, *Of the Conduct of the Understanding*, ed. Francis W. Garforth (New York: Columbia University Teachers College Press, no. 31, 1966), pp. 106-07.

16 Ibid., p. 104; and Locke, "Of Study," p. 418.

17 Locke, *Conduct*, pp. 107-08.

18 John Locke, "Sic Cogitavit de Intellectu humano Jo: Locke an 1671: Intellectus humanus cum cognitionis certitudine, et assensus firmitae," in *An Early Draft of Locke's Essay: Together with Excerpts from His Journals*, ed. R. I. Aaron and Jocelyn Gibb (Oxford: Clarendon Press, 1936), p. 57.

19 John Locke, *An Essay Concerning Human Understanding*, ed. John W. Yolton, 2 vols. (New York: Dutton, Everyman's Library, 1971), 2:258-59.

20 Robert S. Bosher, *The Making of the Restoration Settlement: The Influence of the Laudians, 1649-1662* (London: Dacre Press, 1951), p. 12. See also Paul S. Seaver, *The Puritan*

Lectureships: The Politics of Religious Dissent, 1560-1662 (Palo Alto: Stanford University Press, 1970).

21 Gilbert Burnet, *Bishop Burnet's History of His Own Time*, [ed. Sir Thomas Burnet], 8 vols. (Edinburgh: printed by Hamilton, Balfour, and Neill, 1753), 2:396-97, quoted in *BIOGRAPHIA Britannica: or, the Lives of the Most Eminent Persons Who have flourished in Great Britain and Ireland, From the earliest Ages down to the present Times; Collected from the best Authorities both Printed and Manuscript And digested in the Manner of Mr. Bayle's Historical and Critical Dictionary*, 7 vols (London: Printed for W. Meadows, J. Walthoe, T. Osborne, Dr. Browne, H. Whitridge, C. Hitch and L. Hawes, R. Manby, J. & R. Tonson, John Rivington, C. Bathurst, J. Rivington and J. Fletcher, L. Davis and C. Reymers, J. Ward, R. Baldwin, W. Johnston, J. Richardson, T. Longman, P. Davey and B. Law, T. Kearsley, and M. Cooper, 1747-66), 5:3312.

22 John Pearson, *Opera Posthuma Chronologica, &c. viz. De Serie et Successione Primorum Romae Episcoporum Dissertationes Duae: Quibus praefiguntur Annales Paulini, et Lectiones in Acta Apostolorum. Singula Praelo tradidit, edenda curavit & Dissertationis novis Additionibus auxit H. Dodwellus, A.M. Dubliniensis: Cujus etiam accessit De eadem Successione usque ad Annales Cl. Cestriensis Cyprianicos Dissertatio Singularis* (London: Typis S. Roycroft, L. L. Oriental. Typographi Regis; Impensis R. Clavell, & B. Tooke, in Coemeterio D. Pauli, 1688), cited by Frank Manuel, *Isaac Newton, Historian* (Cambridge: Harvard University Press, The Belknap Press, 1963), p. 266, n. 3.

23 John Pearson, *Vindiciae Epistolarum S. Ignatii. . . Accesserunt Isaaci Vossii Epistolae Duae Adversus David Blondellum* (Cambridge: Typis Joann Hayes, Prostant Londini, apud Guil. Wells & Rob. Scott, ad Insignia Principis in vico Little Britain dicto, 1672).

24 Burnet, *History*, 2:396-97, quoted in *BIOGRAPHIA Britannica*, 5:3312.

25 John Pearson, *An Exposition of the Creed*, 5th ed. (London: Printed by J. M. for John Williams, at the Crown in St. Paul's Church-yard, 1683), p. A.

26 An Ordinance for the more effectual putting in Execution the *Directory for Publique Worship* in all Parish-Churches and Chappels within the Kingdom of England and Wales, 26 August 1645, *Acts and Ordinances of the Interregnum, 1642-1660*, comp. and ed. C. H. Firth and R. S. Rait, 3 vols. (London: H. M. Stationery Office, 1911), 1:755-757.

27 Bosher, *Restoration Settlement*, p. 30.

28 For Cromwell's leniency in the early 1650's toward those inclined toward episcopacy, see Bosher, *Restoration Settlement*, pp. 9-13, 40-41. For the Laudian view of the new order in Church and State, see Bosher, *Restoration Settlement*, pp. 14, 32-39. An account of the Prayer Book controversy of 1653 can be found in Bosher, *Restoration Settlement*, pp. 16-23.

29 Pearson, *Exposition*, 5th ed., p. A2.

30 Pearson, *Exposition*, 5th ed., p. A2.

31 Pearson, *Exposition*, 5th ed., p. A2. Pearson thought innovation—he called it "singularity"—a positive vice in theology. Pearson's fear of "singularity" was hardly singular. His contemporaries shared his prejudice. James Arderne, a cleric and a member like Pearson of the Royal Society, thought all error was "modern and an upstart." James Arderne, *A Sermon Preached at the Visitation of The Right Reverend Father in God John, Lord Bishop of Chester at Chester* (London: Printed for H. Brome, at the Gun at the West-end of St. Pauls Church, 1677), p. 9.

32 Pearson, *Exposition*, 5th ed., p. A2.

33 J. N. D. Kelly, *Early Christian Creeds*, 3rd. ed. (New York: David McKay Company Inc., 1972), p. 5. The two works which Kelly cites are Gerardus Joannes Vossius, *Dissertationes Tres de Tribus Symbolis, Apostolico, Athanasiano, et Constantinopolitano* (Amsterdam: apud Ioannem Blaeu, 1642), and James Ussher, *De Romanae Ecclesiae Symbolo Apostolico Vetere Aliisque Fidei formulis, tum ab Occidentalibus tum ab Orientalibus, in prima Catechesi & Baptismo proponi solitis, Diatriba* (Londoni: Execudebat T. Ratcliffe & E. Mattershed, impensis Johannis Stafford, 1647).

34 Pearson, *Exposition*, 5th ed., p. A2.

35 Pearson, *Exposition*, 5th ed., p. A2.

36 Pearson, *Exposition*, 5th ed., p. A6.

37 Pearson, *Exposition*, 5th ed., p. 2.

38 See Jerome A. Shaffer, *Philosophy of Mind* (Englewood Cliffs, N.J.: Prentice-Hall, Inc., 1968), p. 80.

39 John Norris, *An Account of Reason and Faith: In Relation to the Mysteries of Christianity* (London: Printed for S. Manship, at the Ship near the Royal Exchange in Cornhill, 1697), p. 98.

40 John Tillotson, *A Seasonable Vindication of the B. Trinity. Being an Answer to this Question, "Why do you believe the Doctrine of the Trinity?" Collected from the Works of the Most Reverend, Dr. John Tillotson, Late Lord Archbishop of Canterbury. And the Right Reverend, Dr. Edward Stillingfleet Now Lord Bishop of Worcester* (London: Printed for B. Aylmer, at the Three Pigeons against the Royal Exchange in Cornhill, 1697), p. 5.

41 Isaac Barrow, *A Brief Exposition on the Creed, the Lord's Prayer, and Ten Commandments. To which is added The Doctrine of the Sacraments* (London: Printed by J. H. for Brabazon Aylmer, at the Three Pigeons, over against the Royal Exchange in Cornhill, 1697), p. 13.

42 Ibid.

43 Norris, *Reason and Faith*, p. 61.

44 Norris, *Reason and Faith*, p. 61.

45 John Pearson, *An Exposition of the Creed*, 6th ed. (London: Printed by J. M. for J. Williams, and are to be Sold by Tho. Sawbridge at the Three Flower-de-Luces in Little Britain, 1692), p. 2.

46 Both terms derived from Aristotle's theory that every event could be explained in four ways. Aristotle called these four modes of explanation the four causes. Pearson called them objects. See Aristotle *Physics* 2. 3. 194b16-39; and G. E. R. Lloyd, *Aristotle: The Growth and Structure of His Thought* (Cambridge: Cambridge University Press, 1968), pp. 57-65.

47 Pearson, *Exposition*, 5th ed., p. 2.

48 Pearson, *Exposition*, 6th ed., p. 3. Pearson assumes that the "connexion" is causal. He writes on page 3:

> "For being every natural cause actually applied doth necessarily produce its own natural effect, and every natural effect wholly dependeth upon, and absolutely presupposeth, its own proper cause; therefore there must be an immediate connexion between the cause and its effect. From whence it follows that, if the connexion be once clearly perceived, the effect will be known in the cause, and the cause by the effect. And by these ways, proceeding from principles evidently

known by consequences certainly concluding, we come to the knowledge of propositions in Mathematicks, and conclusions in other Sciences: . . ."

49 Pearson, *Exposition*, 5th ed., p. 3.

50 Bernard Arthur Owen Williams, "Deciding to Believe," *Problems of the Self: Philosophical Papers 1956-1972* (Cambridge: Cambridge University Press, 1973), p. 146.

51 The temptation at this juncture is to give up the struggle for comprehension and attribute our inability to understand Pearson's definition to a radical break in the meaning of belief. The possibility is remote that the concept of belief has changed so radically since the seventeenth century that its meaning bears no connection with its meaning today. Nevertheless, the possibility exists and its supposition in this instance would account for the incongruity of Pearson's definition. It would explain why the seventeenth century found Pearson's definition informative, when we find it uninformative, and it would do so without dismissing their assessment, thereby permitting us to believe what we are unable to deny while extending the same courtesy to the seventeenth century.

Such charity is commendable, but its price is too high. Were we to assume this hypothesis, we would have to sacrifice any hope of adducing a meaning for belief for Pearson and his contemporaries. What "belief" would mean for Pearson and his contemporaries would be untranslatable because nothing in the present meaning of the word would correspond to its meaning in the seventeenth century.

What evidence do we have that our failure to comprehend the definition is inevitable? Experience would lead us to expect the opposite. Pearson's *Exposition* contains a wealth of sentences whose meanings we do comprehend. Even those that escape us are not such as require incommensurability to explain them. If Pearson's definition were untranslatable, we would be forced to question our comprehension of every sentence he ever wrote and to wonder whether his language was totally commensurate with our own. Such doubts seem far less reasonable than assuming that a meaning still exists that would make his definition intelligible.

To this argument from expectation we add another. Not only is it unreasonable to expect the definition to be untranslatable, no evidence exists that could prove that it is. Incommensurability as an hypothesis is unfalsifiable. There are no circumstances where its supposition would be incorrect. Suppose we adduced a meaning for Pearson's definition that was both intelligible and informative. Would this refute the hypothesis that Pearson's definition was incommensurable with our own? No it would not, since the definition could mean something entirely different given another set of assumptions about the meaning of its terms. The fact that we made sense of Pearson's definition would not eliminate the possibility that his definition was the product of an alternative conceptual scheme whose contents bear no relation to the contents of our own. Yet if our ability to make sense of Pearson's definition is no evidence against its possible incommensurability, then our inability to make sense of his definition is no evidence for its possible incommensurability. Though we cannot rule out the possibility of incommensurability, we lack any reason for seriously entertaining it.

Knowing that incommensurability is not a serious possibility sets us free to explore what Pearson meant by belief without the fear that we are distorting his meaning by interpreting his definition of belief in terms of our own. Drawing parallels is the only way we have to grasp his meaning. There is no other option.

The arguments expressed here owe their inspiration to the views of Donald Davidson. See Donald Davidson, "On the Very Idea of a Conceptual Scheme," *Proceedings and Addresses of the American Philosophical Association* 47 (1973-74):5-20.

52 This example is taken from Thomas Jackson, *The Eternal Truth of Scriptures, and Christian Beleefe, Thereon Wholly Depending, Manifested By It Owne Light. Delivered in Two Bookes Of Commentaries upon the Apostles Creede; The former, Containing the positive grounds of Christian Religion in generall, cleared from all exceptions of Atheists or Infidels. The later,*

Manifesting the Grounds of Reformed Religion to be so firme and sure, that the Romanist cannot oppugne them, but with the utter overthrow of the Romish Church, Religion, and Faith (London: Printed by W. Stansby and are to bee sold by John Budge, at the Great South doore of Pauls, and at Britaines-Bursse, 1613), pp. 8-9. Jackson writes:

> "That the Sunne is bigger than the earth, or that the motion thereof is swifter than any arrowes flight, may bee knowne exactly by a Mathematician; but ordinarie country-men (such as are not rustically wayward) do believe it; evidently and exactly know it they cannot. The *ground* of their beliefe in such a case is *authoritas docentis*. And this authoritie of teachers, or others, upon whose assertions we rely, consisteth partly in a perswasion of the teachers or relators skill in those matters which he teacheth or relateth; and partly in his honestie, fidelitie, or veracitie in his dealings, or sayings... As what country-man is there, but would thinke he might safely sweare, that the earth were an hundred times greater than the Sunne? yet if an Astronomer, (of whose skill he hath had experience in other matters, which he can better discerne) one whom he knew to be a honest plaine dealing man, not accustomed to cog with his friend, should seriously avouch the contrary, *that the Sunne is bigger than the earth*; few country men would be so wayward as not beleeve their friend Astronomer albeit (his authoritie set aside) they had no reason to thinke so, but rather the contrarie. ... [Yet, this belief would be strengthened] if the relators or avouchers could make them conceive any probable reason of the same conclusions, (as if the Astronomer in the mentioned case, could shew, how every body the further it is from us seemeth the lesse, and then declare how many many hundred miles the Sunne is from us) mens minds would be a great deale better satisfied, and this assent or beliefe, which formerly did onely relie upon authoritie, would be much strengthened by this second tie, or holdfast."

53 Pearson, *Exposition*, 5th ed., p. 4.

54 Pearson, *Exposition*, 5th ed., pp. 4-5.

55 Pearson, *Exposition*, 5th ed., p. 7.

56 Pearson, *Exposition*, 5th ed., p. 7.

57 Pearson, *Exposition*, 5th ed., p. 12.

58 Isaac Barrow to John Collins, Easter Eve 1669, *Correspondence of Scientific Men of the Seventeenth Century, Including Letters of Barrow, Flamsteed, Wallis, and Newton, Printed From the Originals in the Collection of the Right Honourable the Earl of Macclesfield*, ed. Stephen Peter Rigaud, 2 vols. (Oxford: Oxford University Press, 1841), 2:71.

59 Abraham Hill, "Some Account of the Life of Dr. Isaac Barrow: To the Rev. Dr. Tillotson, Dean of Canterbury," *The Theological Works of Isaac Barrow, D.D.*, ed. Alexander Napier, 9 vols. (Cambridge: Cambridge University Press, 1859), 1:xlviii. See also Walter Pope, *The life of the right reverend father in God Seth, lord bishop of Salisbury, and chancellor of the most noble order of the garter. With a brief account of Bishop Wilkins, Mr. Lawrence Rooke, Dr. Isaac Barrow, Dr. Tubervile, and others* (London: Printed for William Keblewhite, at the Swan in St. Paul's Church-yard, 1697; reprint ed., ed. J. B. Bamborough, Oxford: Basil Blackwell for the Luttrell Society, 1961), p. 152.

60 Ibid., p. 153.

61 A[braham] H[ill] to John Tillotson, 10 April 1683. Napier published Hill's letter in his edition of Barrow's theological works. This is the source of my citation. See Hill, "Life of Barrow," pp. xlvi-xlvii.

62 Verses addressed to John Tillotson by Isaac Barrow, cited and translated by W. Whewell, "Barrow and His Academical Times As Illustrated in His Latin Works," *The Theological Works of Isaac Barrow, D.D.*, ed. Alexander Napier, 9 vols. (Cambridge: Cambridge University Press, 1859), 9:xxxviii-xxxix.

63 Hill, "Life of Barrow," pp. xlviii-xlix.

64 Whewell, "Barrow," p. xlviii.

65 Pope, *Life*, p. 156.

66 Barrow, *Brief Exposition*, p. 13.

67 Barrow, *Brief Exposition*, p. 13.

68 Barrow, *Brief Exposition*, pp. 13-14.

69 Barrow, *Brief Exposition*, p. 12.

70 Barrow, *Brief Exposition*, p. 12.

71 Barrow, *Brief Exposition*, p. 14.

72 Barrow, *Brief Exposition*, p. 14.

73 Barrow, *Brief Exposition*, p. 14.

74 Barrow, *Brief Exposition*, p. 14.

75 Barrow, *Brief Exposition*, pp. 14-15.

76 Barrow, *Brief Exposition*, p. 15.

77 Barrow, *Brief Exposition*, p. 18.

78 Isaac Barrow, *The Theological Works of Isaac Barrow, D.D.*, ed. Alexander Napier, vol. 7: *An Exposition of the Creed; The Lord's Prayer; The Decalogue; The Doctrine of the Sacraments* (Cambridge: Cambridge University Press, 1859), p. 15.

79 Richard Hooker, *The Folger Library Edition of The Works of Richard Hooker*, W. Speed Hill, gen. ed., vol. 1: *Of the Laws of Ecclesiastical Polity: Preface, Books I-IV*, ed. Georges Edelen (Cambridge: Harvard University Press, The Belknap Press, 1977), p. 152. This quote is taken from the first section of the fourth chapter of the second book of Hooker's *Laws*.

80 Jackson, *Eternal Truth*, pp. 2-3.

81 Jackson, *Eternal Truth*, p. 3.

82 Jackson, *Eternal Truth*, p. 2.

83 Jackson, *Eternal Truth*, p. 2.

84 William Chillingworth, *The Religion of Protestants A Safe way to Salvation. Or, An Answer to a Book Entituled Mercy and Truth, or, Charity maintain'd by Catholiques: Which pretends to prove the Contrary. To which is Added The Apostolical Institution of Episcopacy. As also IX. Sermons, The first Preached before His Majesty King Charles the First, the other Eight upon special and eminent Occasions*, 4th ed. (London: Printed by Andrew Clark, for Richard Chiswell at the Rose and Crown in St. Paul's Church-yard, 1674), p. 250.

85 John Tillotson, "Sermon 218: Of the Nature of Faith in General," in *The Works of Dr. John Tillotson, Late Archbishop of Canterbury*, ed. Thomas Birch, 10 vols. (London: Printed by J.F. Dove, St. John's Square for Richard Priestley, High Holborn, 1820), 9:184.

86 John Tillotson, "Sermon 221: Of the Faith or Persuasion of a Divine Revelation," in *The Works of Dr. John Tillotson, Late Archbishop of Canterbury*, ed. Thomas Birch, 10 vols. (London: Printed by J.F. Dove, St. John's Square for Richard Priestley, High Holborn, 1820), 9:240-41.

87 John Wilkins, *Of the Principles and Duties of Natural Religion: Two Books. To which is added, A Sermon Preached at his Funeral, by William Lloyd, D.D. Dean of Bangor, and Chaplain in Ordinary to His Majesty*, 3rd. ed., (London: Printed for T. Basset, R. Chiswell, and C. Brome, at the George in Fleetstreet, the Rose and Crown in St. Pauls Church-Yard, and the Gun at the West-end of St. Pauls, 1693; facs. reprint ed., with an introduction by Henry G. Van Leeuwen, New York: Johnson Reprint Corp., 1969), p. 9.

88 Barrow, *Brief Exposition*, p. 18.

89 Barrow, *Brief Exposition*, p. 10.

90 Barrow, *Brief Exposition*, p. 10.

91 Thomas Jackson, *The Works of the Reverend and Learned Divine, Thomas Jackson, D.D. Sometime President of Corpus Christi College in Oxon; Such as were, and such as Never before were Printed. . . With the Authors Life, And A Large and Useful Table to the Whole, vol. 1: Justifying Faith: or, The Faith by which the just do live. A Treatise Containing a Description of the Nature, Properties & Conditions of Christian Faith. With A Discovery of mis-perswasions, breeding Presumption or Hypocrisie; and means how Faith may be planted in Unbelievers* (London: Printed by Andrew Clark, for John Martyn, Richard Chiswell, and Joseph Clark; and are to be sold at the Bell, and at the Rose and Crown in St. Paul's Church-yard, and at the Star in Little Britain, 1673), p. 614.

92 Isaac Barrow, "Sermon 2: Of the Virtue and Reasonableness of Faith," in *The Theological Works of Isaac Barrow, D.D.*, ed. Alexander Napier, vol. 5: *Seventeen Sermons on the Creed* (Cambridge: Cambridge University Press, 1859), pp. 77, 80.

93 Ibid., p. 82.

94 Isaac Barrow, *The Usefulness of Mathematical Learning explained and demonstrated: Being Mathematical Lectures Read in the Publick Schools at the University of Cambridge. . . To which is prefixed, The Oratorical Preface of Our Learned Author, spoke before the University on his being elected Lucasian Professor of the Mathematics*, trans. Revd. Mr. John Kirkby, of Egremond in Cumberland (London: Printed for Stephen Austen, at the Angel and Bible in St. Paul's Church-yard, 1734; facs. reprint ed., London: Frank Cass and Company Limited, 1970), pp. 64-65.

95 This paean to the Lord as the great Geometer is inscribed on the flyleaf of Barrow's copy of his edition of *Apollonius*, which, Whewell reports, is preserved in the Archives of the Royal Society of London; cited and translated by Whewell, "Barrow," p. xlv. This edition of *Apollonius* is probably *Archimedis Opera: Apollonii Pergaei Conicorum Libri IIII. Theodosii Sphaerica: Methodo Nova Illustrata, & Succincte Demonstrata*, ed. Is[aac] Barrow (London: Excudebat Guil. Godbid, voeneunt apud Rob. Scott, in vico Little Britain, 1675).

96 Barrow, "Sermon 2," *Sermons on Creed*, p. 77.

97 Brian G. Armstrong, *Calvinism and the Amyraut Heresy* (Madison: The University of Wisconsin Press, 1969), p. 32.

98 Stephen E. Ozment, *The Age of Reform 1250-1550* (New Haven: Yale University Press, 1980), pp. 305-06.

99 Armstrong, *Calvinism*, p. 32.

100 E. J. Dijksterhuis, *The Mechanization of the World Picture*, trans. C. Dikshoorn (Oxford: Clarendon Press, 1961), p. 41.

101 Here Hooker's insight is relevant. He writes in the third section of the seventh chapter of the second book of his *Laws* (p. 177):

> "For whatsoever we believe concerning salvation by Christ, although the scripture be therein the ground of our beliefe; yet the authoritie of man is, if we marke it, the key which openeth the dore of entrance into the knowledge of the scripture. The scripture could not teach us the thinges that are of God, unlesse we did credite men who have taught us that the wordes of scripture doe signifie those things."

102 William Payne, *Learning and Knowledge Recommended to the Scholars of Brentwood School in Essex; in a Sermon Preached at their First Feast, June 29, 1682* (London: Printed for Walter Kettilby at the Bishop's Head in St. Paul's Church-yard, 1682), p. 30.

103 Ibid.

104 Ibid.

– 5 –

THE MYSTERIES OF CHRISTIANITY

Two years before the cleric, Thomas Smith, joined the Royal Society in 1677, he published a sermon which defended the reasonableness of believing mysteries like the Trinity and the Incarnation.[1] His apology is of interest not for the arguments he advances, which were standard scholastic fare, but for the view they give of one cleric's understanding (or lack of understanding) of the new critiques of mysteries that were coming to the fore.

What inspired these new critiques and made them plausible were the changes taking place in traditional assumptions about the certainty of human knowledge. The conviction that human knowledge was absolutely certain and thus different from opinion was increasingly being challenged. The reasons for this were many. The past century and a half since the Protestant Reformation had seen the destruction of many certitudes in religion, politics, and the natural sciences. Their destruction had raised the suspicion, fueled by the revival of Greek skeptical thought, that nothing could be known with a certainty that precluded the possibility of error.

Those who entertained this suspicion placed their knowledge of revelation among the things which could be doubted. Though they were willing to believe in the truth of revelation, they doubted whether its contents could be known with absolute certainty. Such doubts were academic when the contents of revelation appeared possible or likely. They were harder to dismiss when the contents of revelation appeared unlikely or impossible.

Critics played upon these doubts when advancing their critiques of the mysteries of Christianity. They argued in their critiques that if the mysteries of Christianity appeared impossible or unlikely, this was reason to doubt the meaning of those sentences in Scripture which attested to their truth. The deist, John Toland, assured his readers that such was his respect for the credibility of revelation that he would sooner doubt his knowledge of the meaning of its text than ascribe to revelation what he knew to be impossible. Whenever its sense appeared to clash with truths of reason or truths of fact, Toland held that he was ignorant of the meaning of its text. Thus he withheld his assent to the mysteries of Christianity because their very impossibility made him wonder whether Scripture really attested to their truth.[2]

Such critiques persuaded many that their knowledge of these mysteries was not absolutely certain and that the doubts these critics raised did appear to have some warrant. But few were willing to concede that their warrant was sufficient to undermine their faith in the mysteries of Christianity. Robert Boyle asserted that the mysteries of Christianity were not absurd or unintelligible because they contradicted truths which were evident to sense or reason. The truths they contradicted could not claim to be so certain that God's negation of their truth was impossible to conceive.[3]

Yet even Boyle acknowledged that so unfathomable were the mysteries proclaimed by Christianity that a person could not be certain whether they understood their meaning. No person, in Boyle's opinion, was obliged to believe these mysteries with the same firmness of assent that they gave to other truths more plainly revealed in Scripture. This was no affront to the veracity of God if a person was disposed to believe what God revealed and only doubted in this instance whether they truly apprehended what God intended to reveal.[4]

It is the argument of this chapter that Thomas Smith saw no warrant in the doubts these critics raised because Smith was not a skeptic—even a "mitigated skeptic." Knowledge as Smith defined it was absolutely certain. Anything less than absolute certainty was not worthy of the name. Smith assumed that he was certain of the mysteries of Christianity. He saw no reason to doubt his certainty because the mysteries of Christianity appeared impossible or unlikely. He assumed that this was evidence of their inexplicability.

Some credit the Royal Society with promoting the very skepticism which is absent from Smith's thought. They claim its program of research entailed a theory of knowledge that questioned whether knowledge could be absolutely certain. If this hypothesis is correct, it leaves unexplained Smith's subsequent activity in the Royal Society.

Not only did Thomas Smith become a fellow of the Society, he was also published in its unofficial organ, the *Philosophical Transactions*.[5] He firmly grasped the methodology which the Society espoused—its distrust of arguments from authority, its enthusiasm for knowledge both factual and useful, its reliance upon observation and experiment, and its recognition of the tentativeness of all hypotheses. Yet Smith embraced this methodology without embracing the theory of knowledge to which it was supposedly conjoined.

The theory of knowledge that Smith embraced is the subject of the section that immediately follows. It is argued in this section that a person could be critical in their pursuit of natural knowledge without renouncing the possibility of achieving absolute certainty. Succeeding sections will discuss Smith's defense of the mysteries proclaimed by Christianity. His apology will be compared with those of other clerics within the Royal Society. Included in this comparison will be apologies by James Arderne, Seth Ward, Benjamin Laney, John Wallis, and William Payne. This comparison will reveal that with the exception of Wallis and Payne none of the clerics who wrote apologies for the mysteries of Christianity thought to doubt their understanding of the meaning of these mysteries. They assumed they were intelligible despite their inexplicability. Any doubts they may have had about the meaning of these mysteries they assumed could be resolved through philological analysis of their sense in revelation.

Science, Signs, and Skepticism in the Thought of Thomas Smith

Smith's grasp of the methodology espoused by the Royal Society is demonstrated in his sermon, published a full two years before his election to the Society. In this sermon, which was dedicated to Robert Boyle, Smith asserted that natural philosophy had a threefold task. Its first objective was "to discover and make out how things were at first made, and still continued in their being;" its second objective was "to find out their peculiar virtues, whereby they produce such a variety of effects;" and its third objective was to learn from these "virtues" how things could "be altered or improved for the farther use and benefit of mankind."[6]

Smith argued that natural philosophy could not achieve these ends "without descending to severe trials and experiments." To acquiesce "in general observations, derived from weak and slight notice," was the method of the scholastics whose theories he castigated as "confessedly unintelligible, and which instead of explaining nature, do but perplex and confound the understanding." Their theories in Smith's opinion had "nothing to maintain and keep up their credit, but the authority of a name and the immoderate love of antiquity."[7]

Smith concluded his assessment of the "business of philosophy" by expressing that caution which was the hallmark of the science practiced by Boyle and the

Royal Society. Smith assured his readers that whatever system was finally chosen as the correct explanation of nature:

> ...they who have the deepest insight into nature will be forced to confess, they see but a little way, and all that they can pretend to is but conjecture and probability, that when they may seem to arrive at some satisfaction in the order and connexion of things, it is very possible and likely, that things may be made and exert their causalities otherwise, than they suppose, be their fancy never so ingenuous, and their reason never so profound and strong...[8]

The doubts Smith expressed concerning the certainty of natural knowledge should not obscure the fact that his conception of knowledge and its relation to opinion remained essentially scholastic. Although Smith attacked the "scientific" value of scholastic philosophy, he never renounced the ideal of knowledge which "science" represented.

Both he and the scholastics were in fundamental agreement concerning the requirements of "science." Both conceived of science as a deductive system of explanation which was necessarily true. The aim of science was 'knowledge of the reasoned fact.' To know that something was the case was to have 'knowledge of the fact.' To know the reason why was to have knowledge of its cause. Causes gave 'knowledge of the reasoned fact.' Only they could explain why things were the way they were and why they could not be otherwise. Knowledge of causes was the goal of science. Smith expressed this goal when he observed that "we are said to know a thing, when we can sum it up to its first principles, can trace its original and cause, and understand its effects and operations."[9]

Neither Smith nor the scholastics thought "science" hypothetical. Scientific theories were either certain or they were not science. Their truth was guaranteed by the fact of their deduction from self-evident propositions. Smith asserted that scientific theories could not be doubted so long as they were drawn "according to the rule and laws of method from first principles, which are assented to, as soon as they are proposed, and the terms understood."[10]

Smith differed from the scholastics on only one point. The scholastics maintained that natural philosophy was capable of science. Smith thought otherwise. He criticized the scholastics for claiming the certainty of science for the systems of nature they had inherited from the ancients. He denied that their systems contained the "original and cause" of the world or the only explanation for all "its effects and operations." Because the ultimate causes of things were not evident to sense or reason, Smith doubted that any system could carry the guarantee of truth which science represented. A person could have 'knowledge of the fact' about the phenomena of nature, but 'knowledge of the reasoned fact' was not within their grasp. A person could learn through observation and experiment that something was the case, but the reason why it was the case remained a matter for conjecture.

Though Smith questioned the absolute certainty of causal explanations, he never doubted there were reasons for the way things were. He never reduced causation to the constant conjunction of discrete events. He continued to maintain that causes did exist and that their nature could be conjectured from the signs of their effects.

Aristotle had taught this in his logic and in his rhetoric. Aristotle had argued that a person could infer the nearness of the planets from the way they did not twinkle. Their not twinkling was a "sign" of their nearness to the earth. Another example of a "sign" was the observed connection between lactation and parturition. Aristotle had argued that a person could deduce that a woman had been pregnant from the fact that she was lactating. Her lactation was a "sign" that she had recently given birth.[11]

Aristotle denied the name of science to the knowledge which signs provided. To know a thing scientifically was to know the cause or reason that accounted for its truth. Signs merely pointed to their causes; they lacked the power to explain them. The appearance of the planets did not furnish an explanation for their nearness to the earth, nor did the presence of lactation account for a woman's parturition. Both were the effects of causes whose nature was unknown. To use the testimony of signs to infer the distance of the planets or a woman's parturition was to engage in conjecture. No matter how probable or likely, such conjectures remained uncertain.

Smith did not belittle the value of such conjectures. The one article he published in the *Philosophical Transactions* was an explanation for the flow of the currents at the Straits of Gibraltar. He acknowledged his hypothesis was only a conjecture and that its truth would "stand or fall according to the certainty of the *Observations*," which he hoped would be made there.[12] Though he was confident that its truth would eventually be upheld, he harbored no illusions that this would end the debate about the reason for the currents. The reason for the currents would continue to be debated so long as the causes of the currents were not evident to sense or reason. The fact that his hypothesis accounted for the currents was no assurance of its necessity. Observations and experiments could point to its truth, but they were powerless to confirm that its truth was necessary. Without this guarantee, his hypothesis could not claim the certainty of science.

Smith's denial of man's ability to know the causes of natural phenomena did not make him a skeptic—even a "mitigated skeptic." Had Smith been a skeptic, he would have doubted the very phenomena which hypotheses explained. No evidence suggests that he doubted there were currents in the Straits of Gibraltar. He apparently believed that the existence of such currents was absolutely certain. The facts about their flow were objects of knowledge because observation and experiment had made them evident to the senses. Knowing the cause or reason that necessitated their truth was not a prerequisite for being certain of their truth. The fact that their truth was evident to the senses was a sufficient guarantee that their truth was necessary.

Had Smith been a skeptic, he would have questioned this guarantee. He would not have assumed that the obviousness of a fact was an infallible criterion of its truth and certainty. He would have demanded further proof of the infallibility of this criterion before trusting its evidence in matters of knowledge. Such a demand would have been justified given the definition of knowledge to which Smith subscribed. This scholastic definition made absolute certainty a requirement for having knowledge. Truths which lacked this certainty were not objects of knowledge, but objects of opinion.

Skeptics used this requirement to criticize the evidence provided by the senses. Skeptics argued that though the obviousness of a fact was a sign of its truth, this sign could not insure that the fact was certain. The possibility still remained that the fact could be evident and still be false. So long as its truth was capable of doubt, the fact lacked the certainty which knowledge required.

Those who attempted to defend the certainty of empirical knowledge soon found they had no proof which could silence the skeptics. Every proof they offered in defense of its certainty supposed criteria of truth which the skeptics could question. This drove some to doubt their knowledge of an external world since it was based upon evidence whose certainty was suspect. Others escaped the dilemma which the skeptics had raised by abandoning the premise which created the dilemma. They denied that knowledge had to be absolutely certain in order to count as knowledge. As a result of their denial, they no longer had to prove to the satisfaction of the skeptics that empirical evidence was an infallible criterion of truth.

Smith never realized that certainty was a problem. He apparently assumed that if a person were truly certain about the things they claimed to know, then the things they knew were certain and beyond the possibility of error. Error was the product of human fallibility. Error could be avoided if a person took precautions not to mistake their certitude for the certainty of knowledge. It never occurred to Smith to ask whether adequate criteria existed for distinguishing mere certitude from the certainty of knowledge. He apparently assumed that true certainty was obvious to anyone who experienced it.

Reason, Faith, and Mystery in the Theology of Thomas Smith

That Smith was not a skeptic accounts for his lack of appreciation for the reasons which led his opponents to doubt the mysteries of religion. Their doubt appeared unwarranted given his framework of assumptions about the nature of knowledge and its relation to opinion. He failed to comprehend that the arguments of his opponents were built upon assumptions that contrasted with his own.

Smith, as we have seen, admitted the possibility of achieving absolute certainty. His opponents denied it. Smith held that knowledge was different from opinion because knowledge was certain and opinion was not. His opponents, on the contrary, thought knowledge and opinion both equally uncertain. They acknowledged the two were different, but they did not follow Smith in construing this difference as a difference in kind. They preferred to think of knowledge as a form of opinion. If knowledge could be said to differ from opinion, the source of its distinction was its greater probability, not its absolute certainty. Thus Smith's opponents shared the conviction that no object was so evident that its falsehood was inconceivable. They also shared the conviction that no opinion was so privileged that its truth could not be questioned.

Smith claimed such a privilege for the mysteries which God revealed. He had trouble understanding why their inexplicability should pose an obstacle to their assent when their truth was guaranteed by the fact of God's approval. Smith complained that his opponents demanded proof where none was needed. They would not assent to mysteries upon the word of God alone, but demanded corroboration from their senses and their reason. Smith thought they were acting unreasonably in spurning God's assurances for the judgment of their faculties.

He assumed "the *command of God* to believe the Testimony *that he has given of his Son*, and the revelations of his will" was sufficient "to move us to assent to the truth of them, though the matter of them be so far above our comprehension."[13]

Such was his regard for the credibility of revelation that he refused to believe that rejection of its witness could ever be justified upon rational grounds. Those who rejected its witness because the things which it attested appeared improbable or unlikely were either deceiving themselves or others if they thought this specious reason could excuse their unbelief. Their demand for further evidence as a price for their assent was tantamount to asking God to tailor His truth to their opinion. Referring to his opponents, Smith observed,

> But all their objections, how plausible soever, must at last resolve into obstinacy and pride; They fancy things must be, and are, as they would have them, or else they cannot be at all: They vainly suppose themselves able to search into the depths of all divine and human knowledge and being prepossessed with this conceit, they grow peevish and angry because the *Christian Religion* proposes things to their belief, which they cannot grasp, and are too big for their understanding; and rather than forego this beloved *Principle*, they will destroy the *Fundamentals* of *Christianity*,...[14]

Thus, Smith dismissed his opponents and the sincerity of their objections. It was his opinion that "the mysteries of Faith do not so much trouble these men, as the severity of its commands." He suspected their only motive in voicing their dissent was the sinfulness of their lives, which would stand convicted if Christianity were true. Since "natural conscience and an ordinary reflexion upon the works of nature" would not permit them to deny a "Creatour of the World," they sought to escape His jurisdiction by denying that the world had a "Judg and Governour."[15] Smith wrote of his opponents:

> ...They it seems are too witty (for so they call their boldness and want of judgment) either to understand or embrace the principles of Christianity; but their ill lives shew, that were they as clear as the principles of *Geometry*, so long as a strict and holy life is as necessary and essential to the being of a *Christian*, as a right and sound faith, they would except and cavil at them, and at last reject them; and if the *Gospel be hid*, be esteemed after so many clear and undoubted revelations, after such evident proofs and convictions, an obscure and incredible doctrine...; it is

> only so to such desperate and obstinate wretches, whom reason it self cannot satisfie,...[16]

Smith did not rest his case against them upon this ad hominem argument alone. Both here and elsewhere, he examined their arguments against the mysteries of Christianity and labored to show their weakness for the benefit of his readers. Smith contended that his opponents were destroying the nature of "belief" when they refused to assent to mysteries upon the authority of revelation. Smith explained:

> To believe in general, in the proper notion of it, is to assent to things upon the discovery and attestation of others, which are not evident and apparent of themselves; that is, when I have no demonstrative or sensible knowledg of things, I admit and judge them to be true, not because I either saw them, and can assure myself of them by any of my other senses, or because they are so evident to my reason, that I must needs embrace them, as a principle or conclusion in Philosophy, but because I have received them from another, who informs me and gives me this account of them, for whose sake I assent to them as real and certain.[17]

It would appear from the above quotation that Smith's definition of belief was the same as John Pearson's. Like Pearson, he made belief a species of assent. He also stipulated like Pearson that no assent could properly be termed belief if the reason why it was given were not the assurance of some authority that its object was true. An assent upon any other motive than the authority's guarantee was not belief, but knowledge or opinion. To believe someone because one shared their opinion was not, according to this definition, to believe them, but to agree with them. Likewise, to believe a proposition for any other reason than the credibility of the person who attested to its truth, was not to believe it, but to agree with it.

Belief implied faith in the credibility of the person whose endorsement of a truth was the reason it was believed. Such faith was reasonable if there were grounds for presuming that the person knew whereof they spoke and spoke according to their knowledge. These grounds could not be taken from the things which they endorsed or their endorsement would cease to function as the cause of the assent. These grounds had to be taken from the person's history of credibility about matters which could be verified. An induction could then be made on the basis of their record whether it was reasonable to believe them in the present instance. This induction was unnecessary in the case of divine authority since it was God's very nature to be omniscient and veracious. Whatever God attested, no matter how unlikely, was credible and certain by reason of His approval.[18]

Thus, "to call for evidence and demonstration in things proposed to be believed," was, in Smith's estimation, "to confound different assents of the mind." The result was "to turn *Religion* into Science" and to deny "the truth of History, and Tradition, and Revelation" because the things which they attested were neither evident to sense nor accountable to reason.[19] Those who restricted their assents to these grounds alone left no room for the certainty which God's approval could provide. Smith was convinced that those who discounted all authority when according their assents would eventually "fall into *Scepticism*, and doubt whether any thing be certain, but what we see and can prove and represent by a Scheme, and at last question whether our Sense, and what we call our Reason do not deceive us." Their mistake was making sense and demonstration the only grounds of certainty. Though the "grand articles of Religion" were neither evident nor clear, God's approval made them as certain "as *Propositions in Metaphysicks* or *Theorems in Geometry.*"[20]

Smith acknowledged that God's approval did not give the believer a scientific understanding of the "grand articles of Religion." Only God could know the reason why these articles were true. No account could be given of the necessity of their truth because "*Supernatural Truths*," unlike the principles of science, were incapable of proofs "derived from nature or sense."[21] Although the reason for their truth was as much a mystery as the reason for the currents at the Straits of Gibraltar, their truth was just as certain as the existence of those currents. God's assurance that they were true was as firm a guarantee of the necessity of their truth as the guarantee provided by observation and experiment.

God used miracles to convey His assurance. Miracles did not demonstrate the truth of the "grand articles of Religion." Their truth was a matter that had to be taken upon faith. What miracles made apparent was the credibility of the miracle-worker. When Jesus first proclaimed the "grand articles of Religion," he asserted his authority to give testimony in God's name by giving evidence of the power which God invested in his person. Such power was a sign of divine authority. Miracles established his credentials as a prophet sent by God and they signaled God's approval of his person and his teaching.

Miracles were the unmistakable effect of divine intervention in the processes of nature. A miracle pointed to its Cause in the very same way that a natural sign like the appearance of the planets or a woman's lactation pointed to the cause which accounted for its occurrence. Miracles surpassed natural signs in the credibility of their witness. Unlike the testimony of natural signs, the testimony of miracles was certain and unambiguous. There could be no question about the identity of their cause. Natural signs, on the contrary, were ambiguous and uncertain. Many causes could be posited to explain their occurrence. Paleness, for example, was a sign of pregnancy, but it could also be occasioned by anemia or fear. Only observation and experiment could narrow the field of choice by specifying the cause that appeared most probable. Even then, the explanation lacked the certainty of knowledge. Miracles were fortunate not to suffer from this impediment. So great was their effect that the only cause which could ex plain them was the hand of God Almighty.

Smith recognized, however, that the testimony of miracles could not compel assents to the truths which God revealed. The "grand articles of Religion" were not evident to sense or reason, and only truths which were so evident could necessitate assents. No miracle could make apparent what was inevident to sense or reason. Truths inevident to sense or reason remained inevident to sense or reason no matter how many miracles were performed on their behalf. Had God intended to compel assents to the "grand articles of Religion," He would have made their truth apparent. God chose not to do so because He wanted their truth accepted upon His authority alone.

Smith also reasoned that if God's authority ceased to function as the motive for their assent, assents to their truth would cease to constitute a virtue and no longer merit the praise that belonged to everyone who believed. Smith wrote:

> One necessary condition to make any activity capable of reward or commendation is, that it flow from a principle of liberty;... if the truths of *Religion* were in themselves so clear and evident, that we could not help but assent, whether we would or no, if they could be prov'd by arguments, deriv'd from sense or nature, where then would be the *blessedness of Faith* our *Saviour* speaks of, which belongs to those, who *have not seen and yet have believed*?[22]

Smith was careful to point out that faith, though voluntary, was not an arbitrary act, for believing was not an action which a person could perform either at will or upon command. The act of believing was subject to constraints which even God could not ignore. No person could believe what God commanded them to believe if the person were not persuaded that the action they were taking was conformable to reason. Thus God had so arranged it that a person could believe what God commanded them to believe if they satisfied themselves that the action was truly rational. The means to this persuasion was available to every person. Both nature and a person's reason provided irrefutable proofs for the being of a God. A person so convinced could deduce for themselves from the very idea of God that the beliefs which God commanded could not be other than true and certain.

Smith nevertheless reported that many failed to be persuaded that belief was truly rational if they could not comprehend how the things which God commanded could possibly be the case. If the things which God commanded ran counter to their experience, they suspended their belief rather than acknowledge the possibility that they were "ignorant of several circumstances" which would render their belief both "probable and easie." Such persons, in Smith's opinion, were being inconsistent. They believed many things "upon the bare testimony of men" which they would not have conceived were possible had they never been informed.[23] If they were willing to believe such things upon the authority of men, what reason did they have for rejecting God's authority? Was God's testimony less reliable than the testimony of men?

Smith also pointed out that those who doubted the rationality of believing what God commanded never used the same excuse when the object of their assent was evident to their senses. The world was full of mysteries which were evident to their senses. They never thought to doubt these mysteries because they could not comprehend them. Smith observed:

> Men are not to be disputed out of belief of their senses, that there is no such thing as motion, or continuity of parts in extended matter, because of the great difficulties, that attend the conception of them, and things are daily produced and by degrees arrive at the perfection of their being, and perform actions suitable to their respective natures, though *Philosophers* disagree in their opinions, and are dissatisfied one with another, and cannot tell how or in what manner they do all this.[24]

If men were willing to believe such mysteries upon the testimony of their senses, why were they hesitant to believe the mysteries which God revealed? Was

God's testimony less infallible than the testimony of their senses? Smith contended that no one could affirm this and believe that God existed, for God by definition was absolutely credible. God's testimony could not be doubted as a source of information. Those who believed that God existed were obliged to believe His testimony regardless of whether its contents appeared possible or not. Those who claimed they were unable were either mistaken or insincere. What prevented them from believing was their obstinacy and their pride.

Smith suspected that those who pleaded their inability to believe what God commanded thought "nothing should exist, but what we can conceive." They were guilty of making God a slave to their conceptions. They ignored the fact that the world was independent of their ability to conceive it. What was true and what was possible were entirely dependent upon the "will and pleasure of God." Neither took its measure from "those narrow conceptions, we borrow from sense," but "from that Idea" God has "in his divine understanding."[25]

Thus Smith was not surprised that the mysteries which God revealed should confound their expectations of what was true and what was possible. The "imperfections" of human reason made this prospect very likely.[26] Reason's powers of comprehension were not commensurate with those of God. To grasp the extent of the disparity, Smith asked his readers to compare their powers of comprehension with those of other creatures. Smith wrote:

> ...if, as we may well suppose, the creatures, which are below us, being of a different judgment and design, if they cannot dive into our councils, and examine their strength and solidity; if an emmet, for instance, or a worm cannot comprehend a piece of art, as the contrivance and mechanism of a watch, or the use of a sector, or quadrant, much less an intricate probleme of Archimedes, or Pappus, or any such curious Mathematical speculation, because life, and motion, and sense are not suitable faculties, and even reason it self is insufficient, without institution and improvement, and the assistance of art, care, and study:...; it is an intolerable presumption and the effect of pride and conceitedness to think, that our shallow reason should be able to comprehend the things of God, the disproportion between God and our faculties being altogether infinite.[27]

Smith concluded that given the "weakness and insufficiency" of human reason, a person had no choice but to take God at His word.[28] Apart from God's authority, no other grounds existed for awarding an assent since the mysteries which God revealed surpassed the power of human reason to ascertain their truth. Reason could not know how such mysteries were possible, though reason could be certain—thanks to the fact of God's approval—that such mysteries were not impossible. Reason's certainty was founded upon its a-priori knowledge that God would not be God if He proposed "any thing to us, as the object of our faith, that is really impossible in it self, and involves in it a perfect and manifest contradiction."[29] On the basis of this knowledge, reason declared itself convinced that nothing should deter it "from the belief of a thing, if *God* has once revealed it."[30] Any difficulty it encountered in the pages of Holy Writ, Smith dismissed as further evidence of reason's "weakness and insufficiency" to judge the truths that God revealed. Smith never wondered whether the difficulties it encountered in the pages of Holy Writ were evidence of its ignorance of the meaning of Holy Writ.

Reason and Authority in the Theology of James Arderne

Thomas Smith was not alone in the strategy he adopted to defend the mysteries of Christianity. Other clerics within the Society chose the very same strategy. This strategy assumed that God's assurance—expressed through miracles—was sufficient to persuade a person of the mysteries of Christianity. Those who claimed they were unpersuaded were being inconsistent since they believed many mysteries, both in nature and natural religion, whose truth and possibility were beyond their comprehension.

James Arderne embraced this strategy in a sermon which he published in 1677. Arderne observed that one of the "fairest Pleas on whose credit a man excepts against what is well established" is "that what has hitherto been receiv'd in the Church, is contrary to the Rules of Reason." Arderne conceded that reason was "very serviceable" in matters of religion and that "it could never find so creditable employment as in defending that of Christians," yet he was careful to point out that its role was one of "Service" and not "Authority." Arderne argued that reason could not legislate in matters of religion because "the Laws of God do not so doubtfully depend upon the sentence of our faculties, nor is the Scripture a fair proposal only that may probably be received, if we approve the matter of it."[31] Arderne wrote:

> All the power of debate, is this, to examine the miracles wrought in confirmation of our faith, and so to consider whether these, and consequently the Gospel which

> accompanies 'em, can proceed from any one but the most High God; and after this hearing is over, reason changes its person, it lays down all its Ensigns of Jurisdiction, descends from its Judgment-seat, and does homage to this Vicegerunt come from Heaven, professing an entire obedience to its Authority; and afterwards if it be found that some of these Laws so Divinely authorized shall appear capable of two meanings, or none that is plain to us, reason is not to take this advantage for an excuse to restore it self to its former power, but it must go to consult with those who convers'd familiarly face to face with the Publishers of this Law, or with their nearest Successors, to know what interpretations they receiv'd of this dark saying, or what Doctrines were deliver'd them nearest the possible meaning of this place, and so it believes according to this true report.[32]

Arderne asserted that if reason did any more than this, it became a "Contriver" and not an "Instrument" in matters of religion, thereby setting its authority against the authority of God. Arderne contended that reason had "no right to take the Doctrines [which God delivers] asunder, and to make an exact search whether these agree to those apprehensions of things which flow from the alone light of Nature." To do so was to question God's omniscience and veracity—two attributes of God that reason could not deny. Thus, Arderne was convinced that reason had no choice but to accept what God declared.[33] Arderne concluded:

> ...he who will attribute to reason an employment higher than this, must be requested to tell us, how by reason he arrived at the belief of the Trinity, and of other mysteries of the Christian Faith. If he say, that he has as distinct a conception of these, as of any self-evident proposition, we congratulate his early attainments and his being already wrapt up into Heaven; but if his free answer prove, as we may well suspect, quite contrary, that for his part he does not at all believe these things, because he cannot clearly understand 'em, and that the Christian Religion has nothing in it, but what is plain and easie to conceive, it will be expected in the next place he shew, after these incomprehensible propositions are laid aside, what so difficult truths there are in our Religion, more than were in the world before; that it should be worthy of God to make a man on purpose, in such a wonderful manner to preach them, and to communicate to him his own power to do those mighty works, that men might believe on this score that which mankind believ'd long afore by the light of Nature.[34]

Seth Ward and the Mysteries of Nature

Like Arderne, Seth Ward assumed that God's authority was sufficient to persuade a person to believe the mysteries of Christianity. He too attacked the notion that a person's ability to believe such mysteries was dependent upon their being "plain and easie to conceive." Those who held to such a notion were suffering in Ward's opinion, from "an Impotency to distinguish betwixt... Truth and Clearness, Obscurity and Absurdity, Being inconceivable and [being] incredible, being incomprehensible and being unaccountable."[35] Ward conceded that the mysteries of Christianity were inconceivable and obscure, but he denied their lack of clearness made them incredible or absurd. Those who held them to be so, subscribed to the opinion "that whatsoever is Inexplicable, Incomprehensible, Unintelligible is (by Philosophers at least and Wise men) to be accounted incredible, fantastical, fictitious, and so to be despised."[36] Ward censured their opinion and claimed that it obliged them to doubt the truth of many mysteries whose evidence was certain. To illustrate his point, Ward drew upon his knowledge of the field of mathematics. Ward observed:

> The most pervious, most clear, and comprehensible of all humane Sciences are the pure and unmixed Mathematicks, yet even in Geometry and Arithmetick how many things are forceably concluded to be true, which are inexplicable, unimaginable, incomprehensible? I shall instance in a very few; So few as need not an Apology to the unlearned in those Sciences: So trivial, as not to admit of an Explication to the Learned.
>
> That the least imaginable space should be equal to another (upon some Base of the same Altitude) whose sides are protracted *in infinitum*, or a Finite greater than an Infinite. The equality of all Circular Angles of Contact. The everlasting approximation and impossible concourse of Asymptots. The Assertions of surd and irrational Quantities, &c. are undeniably demonstrated to be *true*, yet all these (and many more) are *inexplicable, incomprehensible, unintelligible*.[37]

Ward asked those who denied the "Mystery of the Trinity and Unity of the Godhead" whether they clearly understood or fully comprehended the "Affections of Unity and Trinity in Numbers." Were they ready to dismiss the "infinite potential roots of the Number three" because they could not comprehend them? Could they offer an explanation for how the "infinite ascending powers and descending roots" of the number one were "all of them equal, or rather one and the same among themselves?"[38] Ward challenged their assents to mysteries such as these. If these mysteries were as impenetrable as the mysteries of Christianity, what reason could they give for their ability to believe them?

To show that they were prejudiced against the mysteries of Christianity, Ward cited other mysteries that they were guilty of believing. Ward asked them to consider whether the union of the soul and body was any less a mystery than the Incarnation. Were they able to inform him "by what bands and ligaments" the two were joined together or how "Communication betwixt them" could possibly occur? Ward observed that those who followed Thomas Hobbes and made the soul corporeal only deepened the very mystery they were attempting to explain. No appeal to the random movements of "senseless Atoms" or "stupid matter" could ever hope to explicate how the soul was able to sense, imagine, or remember.[39]

Another thing which they believed notwithstanding its great mystery was the phenomenon of "Local Motion." Citing Zeno's paradox of Achilles and the tortoise, Ward demanded that they tell him why they believed in local motion when they knew from Zeno's paradox "that Local Motion was a thing impossible." Ward observed that like the legendary auditor who, hearing Zeno argue, refuted the philosopher by walking back and forth before him, they believed in local motion because it was evident to their senses. Its inexplicability posed no bar to their assent.[40]

Thus Ward concluded that they were prejudiced against the mysteries of Christianity because the mysteries of Christianity, unlike the mysteries of nature, were neither evident to sense nor apparent to human reason. Ward contended that Christ was hardly "so severe upon the understanding of his followers, as to exact a Belief without a sufficient proof and Demonstration of the truth of that which he delivered."[41] The proof Christ offered was neither an a-priori argument "from the common Principles of Intelligence" nor an inductive generalization from the "Phenomena of Nature." Such proofs Christ knew to be "impossible to perform, and senseless to require." Miracles, signs, and wonders were the proof Christ offered. Being "Novel and Anomalous," they were "purposely designed" to communicate God's endorsement of Christ's person and his teachings. Their proof in Ward's opinion was cogent and sufficient and the only kind of proof that revelation would allow. Those who sought to free themselves from their reliance upon such evidence were hankering after proofs both impossible and improper.[42]

Benjamin Laney and the Uses of Skepticism

Benjamin Laney shared Ward's confidence in the certainty of revelation, but he also recognized that a person could not be moved to believe in revelation if that person were convinced that its testimony contradicted what they knew to be the case. This put Laney in a quandary, for he knew he could not prove that their knowledge was in error except by demonstrating the truth of revelation. Such a proof in his opinion was both impossible and illegitimate. Like Smith, Arderne, and Ward, Laney subscribed to the opinion that assents to revelation were "the object and imployment of Faith, and not of Reasoning."[43] He could not condone a proof—had he been able to invent one—that removed God's authority as the motive in such assents.

Yet Laney also understood that God's authority was not persuasive to a person who was certain that revelation was untrue. To persuade them to believe in the truth of revelation, Laney had to make them see that their knowledge of its falsehood was not as certain as they supposed. If their knowledge could be shaken, Laney was confident that they would see that God's opinion of revelation was far more credible than their own.

Laney conceded that no opinion, however credible its source, could persuade them to believe what was contrary to their knowledge. Belief, according to Laney, was the privilege "to see with other mens eyes, as knowledg is with our own." Laney reasoned from this analogy that a person who knew the contrary was "not bound to believe others, for he can see with his own eyes, why shall he be tied to see with other mens?"[44]

Yet Laney warned them not to assume that their knowledge could excuse them from believing revelation if their knowledge was merely negative and not "a *positive* knowing the contrary." To refuse to believe a thing unless one knew or understood it was to leave "no place for Faith at all, which is benefit to see by others."[45] Laney also stipulated that this "*positive* knowing the contrary" must be absolutely certain to excuse them from believing. Laney wrote:

> And for that positive knowledg which discharges us of believing others that we be not mistaken, it is not every conjectural or probable perswasion will do it, but certain knowledg: and when that is, we may safely learn from the Schools; Ubi non est formido contrarii, after diligent search and enquiry, wherein there remains no scruple, doubt, or fear of the contrary, when the understanding is fixed, we are said to be certain.[46]

Since only knowledge which was certain could excuse them from believing, Laney asked them to consider whether the knowledge they opposed to the truth of revelation was really knowledge and not opinion. Laney pointed to the difference between the certainty of things as they are in themselves and the certainty of things as they "come to be represented to and entertained by the Understanding." The latter sort of certainty was not immune to "mistakes and errours." Those who justified their knowledge by the latter sort of certainty confused "Truth in the thing" with "Truth in the apprehension only."[47] Their so-called knowledge could not excuse them from believing revelation because their knowledge lacked the certainty which knowledge claimed to have. Unless they were able to give a reason that would demonstrate its truth, its certainty to them was hardly a reason to prefer it to the truth of revelation. Laney wrote:

> A light in a Candlestick inlightens every one alike that comes into the room. If a light be brought to us which none can see but he that brings it, it may very well be suspected to be none of Gods lights, because it will not stand in a Candlestick, so as to be seen by others. And many such these be, who must be allowed to hold special intelligence with God, and by private illumination from the Spirit, see clearer and farther into the darkest Mysteries than any of the Doctors and Rabbies.
>
> They put me in minde of some, though not the best sort of Creatures, yet in the dark see better than others, as the Owle, the Cat, and the Bat; of which I have known Philosophers give this reason: That rayes of light do naturally stream from their eyes, by which the Medium and Aire about is enlightened; This at best, if there be any such, is a light which none sees but them. Now if those illuminated seers be such that their light is not Gods Word should be in a Candlestick, where it may be seen of all, it is indeed no better then Owles light that shines onely out of their own eyes.[48]

Laney further noted that those who confused "Truth in the thing" with "Truth in the apprehension only" were also liable to do so when interpreting revelation. By interpreting revelation in the light of their opinions, they made the Scriptures merely a mirror to reflect their own opinions.[49] Laney wrote:

> Truly none that I know, but this still, the Scriptures are the only infallible rule. But how! Not left loose to the prejudices and fancies of every man, for then it will fall out, as with those that look in a Glass, in which every one sees his own face, though not anothers; the reason is because he brings his face to the Glass, not

> because it was there before. So every Sect sees the face of his own Religion on the Scripture, not because it was there before, but because his strong fancy and prejudice brought it thither; he thinks he sees that in Scripture, which in truth is onely in his own imagination.[50]

Laney deplored the certitude with which they drew inferences from Scripture which were contrary to tradition and the dictates of the Church. He blamed it on their conviction that inferences drawn from Scripture possessed the certainty of Scripture. Laney countered with the thesis that inferences drawn from Scripture involved the use of human judgment and thus were subject to human error. Laney wrote:

> It [i.e., Scripture] hath all the perfections that a law or rule can have to guide us; yet those perfections are confin'd within the limits and nature of law, to do no more then is proper for a law to do, which is very little without a Judge to apply it. . . . Bring what controversy you will to the lawes, they pronounce nothing wither for the Plaintiff or Defendant. This is the true reason why, though all sects pretend to Scripture, there is yet no end of controversies, because there's no common Judge to end them. And the reason why every sect for all that seems to rest satisfied, they are guided by Scripture, is because they carry their Judge along with them, themselves.[51]

Laney bade his readers to be cautious of the inferences which others drew from Scripture. He urged them to remember that no preacher was infallible when expounding Holy Scripture. Laney observed regarding sermons that "take a Sermon at the best, the most you can make of it is, that it is Gods word onely in a qualified sense, because the Church intends it should be so; and it is the preachers judgment and opinion that it is so; and possibly it may be so indeed."[52] Laney concluded a rule was needed to help a person to determine what God obliged them to believe.

Laney claimed to find this rule in the articles of the Creed. Though this rule was not infallible, it avoided the subjectivity of a rule like private judgment. Laney asserted that the Creed created "no new Religion or new Faith." Its articles were "Articles of Peace,'" not "Articles of Faith." Their function was merely one of "settling unity and consent in matters controverted."[53] Laney wrote in their defense:

> The Forms [of the Creed] are no more but the gathering together some of those Waters which flow over the Scriptures, into a stream, to fit them for ease and use of all. But they say, they take a better course to fetch all from the Fountain. The Fountain indeed is purer; but I see no reason why the Water should be purer in their Pitchers than in the Churches Stream, seeing both claim immediately the same Fountain.[54]

It should be noted in conclusion that Laney's willingness to admit to the possibility of error did not make him a skeptic—even a "mitigated skeptic." Though errors could occur when a person did not distinguish between "Truth in the thing" and "Truth in the apprehension only," Laney apparently maintained that a person could be certain of the things they apprehended if the things they apprehended were evident to sense and reason.

Laney, for example, did not doubt his understanding of the meaning of the Creed. Though there were articles in the Creed whose truth and possibility he was unable to conceive, he did not think to doubt their meaning because he could not apprehend the mysteries they denoted. Laney was convinced that the meaning of these articles was evident to reason once philological analysis had ascertained their meaning. He wrote:

> There are in Religion some things to be known as well as believed, and to these Reason and Discourse is proper; for though the Articles of Faith, and whatsoever shall appear to be contained in Scripture are with out all doubt and reasoning to be received, because God hath revealed them; yet that this or that Article or Proposition is contained in Scripture, is a thing to be known, and lyes within the compasse of sense, we may see whether it be there or no, that is, for the words. And for the meaning of those words we must understand the language in which they are written, the proper import and idiom of the phrase, the force and consequence of the discourse, the coherence and consent it hath with other points better known. We must besides discover the fallacies & inconsequences of those that would obtrude a different sense from that we receive. All these difficulties, though in matter of Religion, are within the conduct of Reason; . . .[55]

Laney assumed that God's authority was sufficient to assure him that the mysteries of Christianity were possible and true. He saw no reason to doubt their meaning unless their meaning was susceptible to rival interpretations. If the evidence for their meaning was clear and unexceptionable, Laney assumed he knew their meaning though he had no way of knowing whether the meaning he apprehended was the one which God intended.

John Wallis and the Argument from Analogy

John Wallis believed, no less than Smith, Arderne, and Laney, that the mysteries of Christianity could have no other ground than the authority of revelation. He affirmed that Christian mysteries like the "Doctrine of the Trinity," "Salvation by Faith in Christ," and the "Resurrection of the Body" were "purely matters of Faith" whose certainty rested solely upon "Divine Testimony."[56] Yet Wallis was willing to use analogies to help a person to conceive how such mysteries were possible. If Wallis believed that God's authority obliged a person to believe, why was he willing to use analogies as a means of persuasion? Did Wallis fear that God's authority was not sufficient to persuade a person of the mysteries of Christianity?

Such questions misconstrue the purpose of his analogies. His analogies did not prove that the mysteries of Christianity were true or even possible; rather they reassured a person that the mysteries of Christianity were not absurd or unintelligible. Wallis used analogies because Wallis understood that a person would not believe in the mysteries of Christianity unless they were persuaded that they truly apprehended what God obliged them to believe.

Wallis considered his analogies a very "great Preparative to the Beleeving of a *Revelation*." Such analogies could not prove independently of revelation that the mysteries of Christianity were either possible or true. Indeed he claimed to be uncertain whether the analogies he drew bore any resemblance to the mysteries they were brought to illustrate. All his analogies served to show was that the sense of revelation could be interpreted in ways that made the mysteries of Christianity "not *Impossible*, not *Improbable*, but very *Likely*."[57]

Wallis, for example, compared the persons of the Trinity to the dimensions of a cube.[58] Wallis did not argue that the comparison was valid. He claimed he did not know this since the Scriptures did not say in what respect the Godhead was one and in what respect the Godhead was three. The Trinity remained a mystery, notwithstanding his comparison, since in Wallis's opinion "we know no more of it than what is revealed in Scripture, [and] where the Scripture is silent, we may be content to be ignorant."[59] What he hoped his comparison proved was that the testimony of Scripture concerning a mystery like the Trinity could be understood in ways that were rational and intelligible.

Such a comparison enabled Wallis to answer the Socinians when they questioned whether Scripture really attested to the Trinity. His comparison removed their reason for doubting the sense of Holy Scripture by showing that the Trinity was not absurd or contradictory. His comparison did not prove that the Trinity represented the true sense of Holy Scripture. Such a proof could not be given since the true sense of Scripture could not be known with absolute certainty. Knowing the Trinity was intelligible reduced the need to raise this question. It allowed Wallis to argue that if the apparent sense of Scripture attested to the Trinity, who were they to doubt its meaning if that meaning was intelligible.

Wallis's use of analogies constitutes a new departure in the apologies we have examined. It reflects a budding awareness of the skeptical problem of meaning. Smith, Arderne, and Ward thought God's authority obliged a person to believe what He commanded. Laney thought so too, though he conceded that God's authority was not persuasive to a person who was certain that God's commands were impossible or contradictory. Laney tried to show using arguments from the skeptics that every person was obliged to believe what God commanded because no person could be certain that the things which God commanded were impossible to believe. Yet Laney, as we have seen, did not broach the problem of meaning. While Wallis did not deny that God's authority obliged a person to believe what He commanded, he also recognized that a person would not concede that they understood God's meaning if the things which God commanded appeared impossible to believe. Such a person would sooner doubt whether they truly apprehended what God obliged them to believe, than believe a revelation that seemed impossible or contradictory.

Though Wallis never said whether a person could assent to the mysteries of Christianity without the benefit of such analogies, he was careful to maintain that such analogies were not sufficient to elicit an assent to the mysteries of Christianity. In showing that these mysteries were "not Impossible, not-Incredible," such analogies merely proved that believing in these mysteries was a more reasonable course of action than denying they were true.[60] Only God's authority had the power to persuade a person that believing in these mysteries was a more reasonable course of action than withholding one's assent.

Wallis likened his analogies to hypotheses in the natural sciences. Both played a similar role in eliciting assents. Natural philosophers used hypotheses as a means of persuasion. They were a means of getting people to pay attention to their proofs when the things they sought to prove appeared absurd or contradictory. Isaac Newton, for example, resorted to hypotheses to help his readers take his meaning when he spoke "abstractedly" of the nature of light and colors. Newton did not think that the hypotheses he employed when discoursing of light and colors had any epistemic value. Indeed he was eager to correct the notion that hypotheses were necessary in a "scientific" discourse. Though hypotheses were sometimes needed for the sake of plausibility, no discourse owed its truth to the fact that it was plausible.[61]

William Payne and the Way of Ideas

The last cleric we will examine, Wallis's colleague, William Payne, also recognized the need to persuade his readers that the mysteries of Christianity were not absurd or unintelligible. He addressed that need in a sermon which was published in 1696.[62] Where Wallis had used analogies to reassure his readers that they understood the meaning of mysteries like the Trinity, Payne attempted to do the same using Locke's theory of knowledge.

Locke made knowledge consist in the perception of the connection between mental objects called "ideas." Ideas mediated between language and reality. Nothing expressed in language could be said to be intelligible unless it signified ideas that the mind could apprehend.[63] Deists like John Toland were quick to seize upon this notion. They claimed the mysteries of Christianity were impossible to conceive because the mind had no idea what these mysteries signified. Since these mysteries signified nothing that the mind could apprehend, they questioned whether these mysteries could be said to be intelligible.[64]

Payne informed his readers that the Trinity was not a mystery because "we have no Idaea, Conception or Knowledge at all of what we do believe concerning it, for then it would be just such a Mystery as it is to those who know nothing at all of it, a Thing perfectly unknown, a Nullity, a Non-Entity, which is no manner of Object of our Faith, and concerning which we can neither affirm or deny any things nor have any Perswasion or Diswasion about it." Payne affirmed that the Trinity was something more than "an empty word, or sound without any meaning, or signification." He assured his readers that "we know very well what is meant by those words of Father, Son, and Holy Ghost, or Spirit, though to connect, and unite those Ideas, and to make such Propositions out of those Terms as are agreeable to Faith may be a difficulty."[65]

Payne acknowledged there were those who claimed the Trinity was unintelligible because they could not form an image "in their Fancy, and Imagination" of the ideas it signified.[66] Payne disputed their assumption that ideas were not intelligible unless the mind could represent them as an image or a picture. He directed their attention to ideas the mind perceived which could not be represented by a picture or an image. Payne observed of these ideas:

> ...we know, and believe, and are well assur'd of a great many things...; as of all Mathematical and Metaphysical Verities about abstracted things; of all Axioms, and Common Notions; of all Demonstrations about Lines and Numbers; of all things not Corporeal, as Time, Motion; the Acts and Operations of our own Minds, such as Understanding, Choosing, Suspending, Doubting, and the like; all which we know by Reason, and can reason about though we can have no Pictures, or imaginary Ideas of them, or any other likenesses of them.[67]

What made the Trinity such a mystery was not the ideas it signified, but the connections God intended these ideas to have. These the mind could not perceive even after God revealed them. Their inconceivability did not make them unintelligible. So long as it was evident that such connections were not impossible, the mind could rest assured that the Trinity was intelligible.

Payne informed his readers that they could learn of these connections in the pages of revelation. He urged his readers to go no further than the sense of revelation when interpreting these connections. They would know they had gone too far if they interpreted these connections in a way that made the Trinity appear absurd or contradictory. Payne warned his readers:

> Yet neither are we so to joyn, and unite Idea's nor to make any such Propositions out of those Terms, as shall be inconsistent with themselves, and destroy, and contradict one another... The doing of this, or endeavouring to do it, has wretchedly perplext our Christian Faith, and made it be called by its Enemies, a Mystery of Nonsense, Absurdity and Contradiction, and exposed it as such to our Heretical Adversaries,...[68]

To illustrate his point, Payne cited 1 John 5:7: "For there are three that bear record in heaven, the Father, the Word, and the Holy Ghost: and these three are one." Payne maintained that a person should not conclude from the meaning of this passage that the Scriptures were advocating a "new way of numbring." Such a literal interpretation of the meaning of the Scriptures was absurd and contradictory. It ignored the possibility that there were other kinds of unity, besides unity of number, that the Godhead might express. It was Payne's opinion that the unity signified in 1 John 5:7 was a unity of essence, not a unity of number. Though the Son and Holy Ghost were one in essence with the Father who was the "Fountain of their Being," such a unity did not preclude their individuation as "real Beings, and subsistences."[69]

Though Payne was confident that he knew that the Trinity was intelligible, he conceded he did not know how the Trinity was possible. Revelation did not tell him why God produced a Son, whether the emanation was necessary or voluntary, how the procession of the Holy Spirit differed from the generation of the Son, or how the Son was able to take flesh while not disturbing the unity of the Godhead. Such mysteries remained for Payne:

> ...Incomprehensible, Unconceivable, and Unaccountable to our Reason, by which I mean only that we can compare 'em to nothing else that we certainly know, that we have no perfect, or absolute likeness of them or Representation of any such thing in our minds from anything else in Nature, but that 'tis a Truth purely known, and manifested by Revelation, joyning together, and connecting such Idea's concerning God, and so presenting them to the mind as united together under one view, and so compounded, and complicated with one another that tho' they do not disagree, nor are contrary to one another yet we see not perfectly their agreement, or connection, but have only an obscure, and general, and confused view, and conception of them, as we have a sight of such Objects, which are too high, or too remote for our eye thoroughly to kind,...[70]

Payne went on to argue that there were mysteries in nature whose inexplicability rivaled that of the Trinity. No one doubted they were true or said they were unintelligible because of their inexplicability. Payne wrote:

> Thus in Physicks, in Mathematicks, in Chymistry, in Astronomy, in Navigation, and other Sciences, how many strange and extraordinary and wonderful things are there which those who are ignorant would hardly believe, and to those who know them they are Incomprehensible... Who would not have thought the Loadstone a very Mysterious thing, and the Pixis Nautica when first found out,

> and who does not think them so still, and who can give a full account of them by striate Particles and the Magnetism of the Earth, which is going very little further than an occult quality. And a great many such occult qualities and unaccountable Appearances there are in Nature, and Phylosophy, which are much like Mysteries in Religion and Revelation, which when our knowledge is improved we shall know better. The Flux and Reflux of the Sea continues to be so still for any Hypothesis that can perfectly solve it. The Circulation of the Blood through the whole Body was not long ago thought to be so, and the Periods of Agues and Feavers is still acknowledged to be one. The believing there were Antipodes was not many Ages ago thought to be Heresy... The Phylosophers Stone has been greatly talked of, and will be acknowledged a Mystery though so many have pretended to it, and even though it were found out. The Power of Numbers, and the performances of Algebra, and many Problems in Mathematics may pass for Magical to a great many, as they have done always to the ignorant; and some of them are unaccountable and Incomprehensible to those who know and can demonstrate them; as the Infinite Division of Numbers in Decimal Fractions; the Cases of asymptotes. How two Lines may be so drawn, as always to approach and come nearer to one another, tho' drawn in infinitum.[71]

Payne complained of the double standard which allowed mysteries in natural religion and condemned them in revealed. He argued this implied that only mysteries in revealed religion were inexplicable to human reason. Such an assumption was incorrect. Payne cited mysteries in natural religion which were just as inexplicable as those in revealed religion. He wrote:

> There are Incomprehensible Truths in Natural Religion as well as in Revelation, and in the Mystery of Christian Faith. God who is known by Nature, and whose Being and Perfections are the great objects of natural Religion, is as Incomprehensible and as much too big for our finite Reason and Faculties fully to Comprehend as any of the Articles of Revealed Religion or Christian Faith. His (autasia) or Self-Existence is such a thing as our Thoughts are wholly lost about when we consider of it, and yet something must be self-existent even by the Atheist hypothesis. So something must be Eternal, and yet let any man try his thoughts about Eternity, Infinite Duration, or Infinite Succession, and see how swallowed up in those deeps of Speculation? Let him conceive if he can Eternity with out Succession, and Immensity with out Extension, and clear himself from that Contradiction, if applyed to any thing but God, that what is past or future is yet

> present, that he now exists, tomorrow, and that a thousand years ago his Duration was the same as now, and nothing is added to it, he may improve these into as puzling and incomprehensible difficulties if he pleases, as any thing about the Trinity.[72]

Payne contended that those who withheld their assent to the mysteries of Christianity because they found them inexplicable were demanding that God reveal what they were able to comprehend. He criticized their arrogance in making this demand. Payne observed that "at this rate Revelation must never go beyond our Natural Knowledge." He accused them of accepting nothing upon the authority of God, but demanding "other reasons" to know these mysteries were true. Such a demand, in Payne's opinion, made "our selves, and not God, Judges of what is true and false."[73] Payne concluded:

> The Divine Power is no more limited to do those things only which we can conceive, than the Divine Veracity to Reveal those things only which we can Comprehend, and yet we are to believe both. If we could conceive fully how the one [i.e. miracles] was done, we might do it our selves; and if we could comprehend perfectly how the others are true, and how those several Ideas are connected and agree together, we might compound them our selves; and so God would not be more Powerful or more Wise than Man is. Our Finite Understanding is no more than our Finite Power to effect every thing that he can. Miracles are as much above us as to Facts, and as much Mysteries as to the Laws of Matter and Motion and the known properties of Bodies, as Revealed Articles are above us as Truths and known principles of Reason, and as to any other comparative Ideas and representations of things which are Familiar to us, and which we are well acquainted with.[74]

Conclusion: The Skeptical Problem of Meaning

Payne's comments above are illustrative of the decline of authority as a reason for believing and the erosion of the fiduciary element from the concept of belief. Both were symptomatic of the rise of skepticism concerning the certainty of human knowledge. Before the rise of skepticism it was commonly assumed that authorities could have knowledge which was absolutely certain. Such a presumption made it unreasonable to question an authority's credibility if the things which they attested appeared improbable or unlikely. A person was obliged to believe what they attested unless the things which they attested were evidently false. The rise of skepticism removed this obligation. If authorities could not have knowledge which was absolutely certain, no person was obliged to believe what they attested if their testimony contradicted what was probable or likely.

Skeptics did not deny that a person was obliged to believe what God attested. Since God was infallible and veracious, God's testimony could not be doubted as a source of information. What skeptics sought to question was a person's understanding of the meaning of God's testimony. Though God's testimony could tell a person what mysteries God held true, it could not tell a person what God meant by these mysteries nor what belief their holding them true might conceivably represent. Skeptics were convinced that a person should conclude that they had misconstrued God's meaning if the mysteries God revealed seemed impossible to believe.

This chapter has suggested that some of the clerics within the Society, namely, Smith, Arderne, Ward, and Laney, did not doubt their understanding of the meaning of revelation because they did not doubt the certainty to which their knowledge could attain. They assumed that they were certain of the meaning of revelation if the meaning of revelation was evident to their reason. They saw no reason to doubt its meaning because the mysteries it attested appeared impossible or unlikely. They assumed that God's authority was sufficient to assure them that the mysteries of revelation were true if inexplicable.

Other clerics did not assume that the meaning of revelation could be known with absolute certainty. Payne and Wallis both acknowledged that a person could find reason to doubt the meaning of revelation if the mysteries God revealed appeared impossible or unlikely. They argued that a person should assume they knew God's meaning if they were able to construe it in a way that was intelligible.

NOTES

1 Tho[mas] Smith, *A Sermon of the Credibility of the Mysteries of the Christian Religion. Preached Before a Learned Audience* (London: Printed by Tho. Roycroft for Ric. Davis Bookseller in Oxford, 1675).

2 [John Toland], *Christianity not Mysterious: or a Treatise Shewing, That there is nothing in the Gospel Contrary to Reason, Not Above it: And that no Christian Doctrine can be properly call'd A Mystery* (London: n. p., 1696; facsimile reprint ed., New York: Garland Publishing, Inc., 1978), pp. 34-35.

3 God was able to negate their truth because their truth was contingent. Had their truth been logically necessary God's negation of their truth would not have been possible to conceive. Boyle observed, for example, that creation ex nihilo was not absurd or unintelligible though its truth affirmed a thing that was physically impossible. So long as its truth remained a logical possibility, creation ex nihilo was possible to conceive. See [Robert Boyle], *Some Considerations about The Reconcileableness of Reason and Religion To Which is annex'd by the Publisher, A Discourse of Mr. Boyle about The Possibility of the Resurrection* (London: Printed by T. N. for H. Herringman, at the Anchor in the Lower Walk of the New Exchange, 1675), pp. 93-94.

4 Ibid., pp. 113-14.

5 Thomas Smith, "A conjecture about an under-Current at the Streights-Mouth read before the Oxford Society, Dec. 21. 1683," *Philosophical Transactions Giving Some Accompt of the Present Under–takings, Studies, and Labours, of the Ingenious in many Considerable Parts of the World* 14 (April 20, 1684): 564-66.

6 Smith, *Credibility of Mysteries*, pp. 16-17.

7 Smith, *Credibility of Mysteries*, p. 17.

8 Smith, *Credibility of Mysteries*, p. 17.

9 Smith, *Credibility of Mysteries*, p. 28. The distinction between 'knowledge of the fact' and 'knowledge of the reasoned fact' is Aristotelian. Aristotle discusses it in the *Posterior Analytics* 1. 13. 78a22-34. For an illuminating account of Aristotle's theory of scientific explanation, see B. A. Brody, "Towards an Aristotelian Theory of Scientific Explanation," *Philosophy of Science* 39 (March 1972):20-31. See also G. E. R. Lloyd, *Aristotle: The Growth and Structure of His Thought* (Cambridge: Cambridge University Press, 1965), pp. 122-25, and W. D. Ross, *Aristotle*, (New York: Meridian Books, Inc., 1959), pp. 44-58, for their discussion of Aristotle's understanding of the nature and goal of "science."

10 Smith, *Credibility of Mysteries*, p. 28. Aristotle argues in the *Posterior Analytics* 1. 2., that the primary premises of any science must be true and immediate, and the ultimate explanation for all the conclusions in that science. Hintikka interprets Aristotle as endowing these premises with existential import. If this import is carried downwards by means of syllogistic reasoning, as Hintikka affirms, then the conclusions which science demonstrates cannot be other than real and necessary. See Jaakko Hintikka, "On the Ingredients of an Aristotelian Science," *Nous* 6 (1972):62-66.

11 Aristotle writes in the *Prior Analytics* that "that which coexists with something else, or before or after whose happening something else has happened, is a sign of that something's having happened or being." See Aristotle *Prior Analytics*, Loeb Classical Library, 2. 27. 70a8-10.

The sign that planets are near the earth because they do not twinkle is used by Aristotle in the *Posterior Analytics* 1. 13. 78a32-40, to illustrate the difference between 'knowledge of the fact' and 'knowledge of the reasoned fact.' Signs provide the former when they are used as middle terms in enthymematic syllogisms. Only demonstrative syllogisms give the latter kind of knowledge since their middle terms explain why the objects they are proving are true and necessary. See Edward H. Madden, "The Enthymeme: Crossroads of Logic, Rhetoric, and Metaphysics," *Philosophical Review* 61 (July 1952):368-73; and Brody, "Scientific Explanation," p. 22.

The sign that a woman has given birth because her breasts are giving milk is cited by Aristotle in the *Prior Analytics* 2. 27. 70a11-39, as an example of a sign whose evidence is conclusive. Such a sign is called by Aristotle a "necessary sign." Necessary signs are irrefutable if they are used in enthymemes that conform to the first figure. Most signs are only probable and can easily be refuted by counter-examples. Thus the sign that a woman is pregnant because she is pale is a probable sign. Aristotle discusses the nature of signs and their relation to enthymemes in the *Rhetoric* 1. 2. 1357b 18; *Rhetoric* 2. 25. 1402b. 8-9; and *Rhetoric* 2. 25. 1403a. 12-14. For an overview of the role of signs in Aristotle's *Rhetoric* and in the Greek rhetorical tradition, see G. A. Kennedy, *The Art of Persuasion in Greece* (Princeton: Princeton University Press, 1963), pp. 99-100; and Friedrich Solmsen, "The Aristotelian Tradition in Ancient Rhetoric," *American Journal of Philology* 62 (January-April 1941):40, 171-72.

12 Smith, "Conjecture," p. 566.

13 Tho[mas] Smith, *A Sermon Concerning the Doctrine, Unity, and Profession of the Christian Faith. Preached before the University of Oxford, To which is added an Appendix concerning the Apostles Creed* (London: Printed for Walter Kettilby, at the Bishop's Head in St. Paul's Church-Yard, 1682), p. 31.

14 Smith, *Credibility of Mysteries*, pp. 11-12.

15 Smith, *Credibility of Mysteries*, p. 9.

16 Smith, *Credibility of Mysteries*, pp. 10-11.

17 Smith, *Credibility of Mysteries*, p. 27.

18 Smith wrote in *Credibility of Mysteries*, p. 41:

> "The Articles of *Faith* carry along with them sufficient motives of *Credibility*, but then these motives must not be fetched from the nature of things themselves, as if they were to be so evident, that our Reason might fully discover their connexion and dependance, but from without; that is, my *Faith* is rightly grounded, and an obligation lies upon me to believe, what is proposed by God, if it be evidenced so to be, by just and rational proofs; and if the authority be certain and infallible: God therefore declaring his Will, and confirming the Revelations he has made of it by his divine Power, this latter is a sufficient proof, and just and rational ground of my Belief;..."

19 Smith, *Credibility of Mysteries*, pp. 28-29. Smith repeated this point in *Christian Faith*, p. 32. Smith contended that if the "Articles of Faith" were restricted to those articles which could be proven by sense and demonstration, "they would cease to be matters of *Faith*, and lose their name, and fall under the comprehension of *Science*."

20 Smith, *Credibility of Mysteries*, p. 29.

21 Smith, *Credibility of Mysteries*, p. 29.

22 Smith, *Credibility of Mysteries*, pp. 33-34.

23 Smith, *Credibility of Mysteries*, pp. 25-26.

24 Smith, *Credibility of Mysteries*, p. 19.

25 Smith, *Credibility of Mysteries*, pp. 18-19.

26 [Vincentio Puccini], *The Life of St. Mary Magdalene of Pazzi, a Carmelite Nunn. Newly translated out of Italian by the Reverend Father Lezin de Sainte Scholastique, Provincial of the Reformed Carmelites of Touraine. At Paris, For Sebat. Cramoisy in St. James's Street, at the Sign of Fame. 1670. And now done out of French: With a preface concerning the nature, causes, concomitants, and consequences of Ecstacy and Rapture, and a brief discourse added about discerning and trying the spirits, whether they be of God*, trans. [Thomas Smith], preface by [Thomas Smith] (London: Randal Taylor, near Stationers Hall, 1687), p. 96.

27 Ibid., p. 97.

28 Smith, *Credibility of Mysteries*, p. 18.

29 Smith, *Credibility of Mysteries*, p. 24.

30 Smith, *Credibility of Mysteries*, p. 39.

31 James Arderne, *A Sermon Preached at the Visitation of The Right Reverend Father in God John, Lord Bishop of Chester at Chester* (London: Printed for H. Brome, at the Gun at the West-end of St. Pauls Church, 1677), pp. 6-7.

32 Ibid., pp. 7-8.

33 Ibid., pp. 6-7.

34 Ibid., p. 8.

35 Seth Ward, *An Apology for The Mysteries of the Gospel. Being A Sermon Preached at White-Hall, Feb. 16 1672/3. Printed by his Majesties special Command* (London: Printed by Andrew Clark, for James Collins, 1673), pp. 14-15.

36 Ibid., p. 24.

37 Ibid., p. 26.

38 Ibid., pp. 26-27.

39 Ibid., pp. 27-28.

40 Ibid., pp. 25-26.

41 Ibid., pp. 34-35.

42 Ibid., pp. 30-31.

43 Benjamin Laney, *A Sermon Preached before His Majesty at Whitehall March 9th. 1661... Published by His Majesties Special Command* (London: Printed by R.N. for Timothy Garthwait, at the Kings-Head in St. Paul's Churchyard, 1662), p. 22.

44 Benjamin Laney, "The Study of Quiet, In two Sermons Upon the same Text, Fitted to give an allay to the Heats of these Unquiet Distemper'd times. In which are particularly conteyn'd necessary Instructions to the Student about Way and Means of attaining the fruit of his Study,

and setting him out of the danger and necessity of seeking it by the New device of Comprehension," in *Five Sermons, Preached before His Majesty at Whitehall. Published Severally by Command. And Now Printed together, tending all to give satisfaction in Certain Points to such who have thereupon endeavoured to unsettle the State and Government of the CHURCH* (London: Printed for Timothy Garthwait, 1669), p. 88. The first sermon was preached before the King on March 12, 1664/5; the second, March 18, 1665/6.

45 Ibid., p. 89.

46 Ibid., p. 89.

47 Ibid., pp. 99-100.

48 Benjamin Laney, *A sermon preached at Whitehall before His Majesty March 27th 1664* (London: Printed for Timothy Garthwait, at Kings-Head in St. Paul's Church-yard, 1665), p. 35.

49 Ibid., p. 12.

50 Ibid., p. 12.

51 Laney, *Sermon Preached March 9th 1661*, p. 25.

52 Laney, *Sermon Preached March 27th 1664*, pp. 9-10.

53 Laney, *Sermon Preached March 27th 1664*, p. 14.

54 Laney, *Sermon Preached March 27th 1664*, p. 16.

55 Laney, *Sermon Preached March 9th 1661*, p. 22.

56 John Wallis, *The Resurrection Asserted: In a Sermon Preached to the University of Oxford, on Easter-Day, 1679* (Oxford: Printed by Hen: Hall for James Good, 1679), p. 24.

57 John Wallis, *The Life of Faith. In Two Sermons to the University of Oxford, At St. Mary's Church there; On the 6th. of January, 1683/4 and June the 29th. following* (London: Printed by James Rawlins, for Thomas Parkhurst: And are to be sold by Amos Curteine, Bookseller in Oxford, 1684), p. 30.

58 See John Wallis, *The Doctrine of the Blessed Trinity Briefly Explained to a Friend* (London: Printed for Tho. Parkhurst, at the Bible and Three Crowns, in Cheapside, 1690), pp. 11-15; and John Wallis, *An Explication and Vindication of the Athanasian Creed, In a Third Letter, Pursuant of Two former, Concerning the Sacred Trinity. Together With a Postscript, in Answer to another Letter* (London: Printed for Tho. Parkhurst, at the Bible and Three Crowns, in Cheapside, 1691), p. 35.

59 Wallis, *Doctrine*, p. 4.

60 Wallis, *Resurrection*, p. 25.

61 Newton to Oldenburg, 7 Dec. 1675, *The Correspondence of Isaac Newton*, ed. H. W. Turnbull, 4 vols. (Cambridge: Published for the Royal Society at the University Press, 1959), 1:363.

62 William Payne, *The Mystery of the Christian Faith and of the Blessed Trinity Vindicated and The Divinity of Christ Proved, In Three Sermons, Preach'd at Westminster-Abbey upon Trinity Sunday, June the 7th and September 21, 1696. By the late Reverend William Payne, D.D. In the Press before his Death, and by himself ordered to be Published* (London: Printed for Richard Cumberland, at the Angel in St. Pauls Church-Yard, 1697).

63 For Locke's theory of ideas, see John Locke, *An Essay Concerning Human Understanding*, ed. John W. Yolton, Everyman's Library, 2 vols. (London: J. M. Dent & Sons Ltd, 1965), particularly bk. 2, chap. 1, where he discusses the nature of ideas; bk. 3, chap. 2, where he discusses how ideas are signified by words; and bk. 4, chaps. 1, 3, and 5, where he discusses the relation of ideas to truth and knowledge.

64 See [Toland], *Christianity not Mysterious*, particularly pp. 27, 34-35, 140.

65 Payne, *Mystery*, pp. 17-18.

66 Payne, *Mystery*, p. 33.

67 Payne, *Mystery*, p. 33.

68 Payne, *Mystery*, p. 18.

69 Payne, *Mystery*, pp. 21-26.

70 Payne, *Mystery*, pp. 29-30.

71 Payne, *Mystery*, pp. 48-50.

72 Payne, *Mystery*, pp. 50-52.

73 Payne, *Mystery*, pp. 60-62.

74 Payne, *Mystery*, pp. 55-56.

– 6 –

CONCLUSION

This study has examined fifty-three clergymen who were members of the Royal Society of London between 1663 and 1687. Many of those elected were men of note who had achieved high office in the Church of England. Chapter 1 disclosed that six clerics were bishops at the time of their election, including one who was archbishop of Canterbury. Another eight became bishops during the course of their membership in the Royal Society. Seven other clerics ministered to fashionable London congregations. Hence it appears in many cases it was social position rather than an aptitude for experimental science that explains their election to the Royal Society.

Another group of clergy had more leisure to pursue scientific interests as ordained academics. If we exclude the one minister who did not conform at the Restoration, we find just under half were exercising no pastoral office at the time of their election. The majority of this group were resident members of Oxford and Cambridge. Seven held endowed chairs or lectureships, twelve were fellows or tutors, and one was a college president. Of the remaining members of this group, five held academic posts in London while four were peripatetic scholars. It is among this academic group that we find the liveliest scientific minds.

It seems likely that the majority of clerics joining the Society in this period first had their interest sparked in science during their years at university. A look at the educational backgrounds of the clerical virtuosi supports this supposition. Forty-five of the fifty-three had M.A.'s from either Oxford or Cambridge. A further three had M.A.'s from institutions outside of England. Only three were entirely products of the Inns of Court—one of the alternatives to an university education. This preponderance is significant because 40 percent of the M.A. course of study was devoted to science.

Even more striking is the number of clerics who proceeded to advanced degrees. Excluding creations, twenty-six clerics attained the D.D., two stopped with a B.D., and three achieved the M.D. Robert G. Frank has argued that the "man who terminated his education at the an advanced level, at the M.A. or beyond, would be much more able to define important problems for himself, to carry out the work necessary to solve them, and to have the inclination and confidence to present his conclusions to his peers for evaluation."[1]

Their academic background sheds new light upon the motives that some of the clergy may have had for joining the Royal Society. Robert G. Frank has suggested that the Royal Society owed its existence and success to a simple demographic fact. London contained the largest concentration of university educated men outside Oxford and Cambridge. Frank theorizes that the foundation of the Royal Society represented an attempt on the part of university men "to continue, indeed to perpetuate, the kind of active association concerning learned topics which they found in their universities."[2] If Frank's theory is correct, it would appear, for example, that Bishops Morley and Dolben and Archbishops Sheldon and Stearne were doing more than following the fashion of the Court when they joined the Royal Society in the middle sixteen-sixties. Though their subsequent inactivity gives no hint of any interest in natural philosophy, it is also the case that during their university careers they resided in colleges which were centers of scientific thought.

We have seen in Chapter 2 that almost half of the clerics who were elected in this period were not active in the Society after the date of their election. Though this is a sizeable amount of deadwood it should not blind us to the presence of some finer grains. We have seen in Chapter 1 that at least twenty-two clerics pursued scientific interests that ranged from agriculture to astronomy. It is significant that the work of ten receives mention in the *Dictionary of Scientific Biography*.

The dominant interest of this study has been less with prosopography than with certain aspects of their thought that shed light upon the relation of religion and science, faith and reason, in the latter seventeenth century. These aspects document and illuminate the increasing rationalism of Restoration theology. Mark Pattison wrote in his seminal essay on eighteenth-century religious thought that rationalism permeated religious thinking in the period. Pattison observed:

> Rationalism was not an anti-Christian sect outside the Church making war against religion. It was a habit of thought ruling all minds, under conditions of which all alike tried to make good the peculiar opinions they might happen to cherish. The Church-man differed from the Socinian, and the Socinian from the Deist, as to the number of articles in his creed; but all alike consented to test their belief by the

> rational evidence for it. Whether given doctrines or miracles were conformable to reason or not was disputed between the defence and the assault; but that all doctrines were to stand or fall by that criterion was not questioned.[3]

We have seen in Chapter 4 that both the clerical virtuosi and their theological opponents were rationalists to the core. Both agreed that their opinions should be put to the test of reason. They only differed in their assessment of the significance of the evidence to be used in such a test. All agreed, for example, that the mysteries of Christianity appeared improbable and unlikely when judged solely in the light of their analogy to experience. Yet the clerical virtuosi thought this evidence was insignificant when weighed against the Scriptures and the light that they provided. They had trouble understanding why Deists and Socinians found such evidence unconvincing. What evidence could be more credible than the testimony of God? Deists and Socinians were equally perplexed. They had trouble understanding how anyone could hold that God was the author of these mysteries when there was nothing in these mysteries that suggested they were true. Would the God of Truth attest to things that appeared to be untrue? Since these mysteries appeared in Scripture, they attributed their presence to corruptions in the text or impositions upon its meaning of a sense it did not have.

Thus each side accused the other of willfully ignoring evidence that was relevant to the case. Such accusations marked a breakdown in the consensus concerning the nature of appropriate reasoning, decisive evidence, and conclusive proof. We have seen in Chapter 3 that this breakdown had its origin in contrasting points of view concerning the certainty of knowledge and the avoidability of error. These contrasting points of view were reflected in the way that the clerical virtuosi defined the meaning of belief. Some defined belief as if authority was the only motive that was relevant in belief. Those who held to this definition did not think to draw a line between reason and authority. They assumed their beliefs were rational though grounded upon authority. According to their reasoning, if authorities could have knowledge which was absolutely certain, it was reasonable to trust their testimony regardless of its probability provided the authorities were credible. Others defined belief without specifying a motive that was proper to the assent. Any motive was appropriate if it purported to show the truth of the thing to be believed. Since they denied that knowledge was absolutely certain, they denied that it was rational to trust authorities any further than the likelihood of their testimony. According to their reasoning, if authorities could not have knowledge which was absolutely certain, then even the most credible authorities were liable to error. No longer could authorities claim their knowledge was so certain that it guaranteed their testimony.

We have seen in Chapter 4 that some of the clerical virtuosi were able to address the criticisms made by Deists and Socinians against the mysteries of Christianity because they shared their opponents' views concerning the certainty of knowledge and the avoidability of error. They acknowledged that these mysteries appeared to be improbable and that it was reasonable to doubt their meaning on the basis of this evidence. They attempted to allay their doubts, not by insisting upon God's authority, but by showing their opponents that these mysteries were not absurd, but intelligible.

NOTES

1 Robert G. Frank, Jr., "Science, Medicine and the Universities of Early Modern England: Background and Sources, Part I," *History of Science* 11 (1973):199.

2 Ibid.

3 Mark Pattison, "Tendencies of Religious Thought in England, 1688-1750," in *Essays and Reviews* (London: John W. Parker and Son, West Strand, 1860; facs. reprint ed., Westmead, Farnborough, Hants: Gregg International Publishers Ltd., 1970), pp. 257-58.

SELECTED BIBLIOGRAPHY

Primary Sources

Arderne, James. *A Sermon Preached at the Visitation of The Right Reverend Father in God John, Lord Bishop of Chester at Chester.* London: Printed for H. Brome, at the Gun at the West-end of St. Pauls Church, 1677.

Aristoteles. *Aristotle's Posterior Analytics.* Translated by Jonathan Barnes. Clarendon Aristotle Series. Oxford: Clarendon Press, 1975.

________. *The "Art" of Rhetoric.* Translated by John Henry Freese. Loeb Classical Library. Cambridge: Harvard University Press, 1957.

________. *Posterior Analytics.* Translated by Hugh Tredennick. In *The Organon II*, pp. 24-261. The Loeb Classical Library. Cambridge: Harvard University Press, 1960.

________. *Prior Analytics.* Translated by Hugh Tredennick. In *The Organon I*, pp. 199-531. The Loeb Classical Library. Cambridge: Harvard University Press, 1938.

Ashe, St. George. "An Account of the Subsiding, or Sinking down of Part of a Hill near Clogher in Ireland. Communicated by the Right Reverend the Lord Bishop of Clogher, F. R. S." *Philosophical Transactions* 28 (1713):267-69.

________. "Extract from the Bishop of Cloyne's Letter, near Youghall, April 2d. 1696." *Philosophical Transactions* 19 (March-May 1696):224.

________. "A new and easy way of demonstrating some propositions in Euclid by the learned Mr. — Ash. a Member of the Philosophical Society of Dublin for promoting natural Knowledge." *Philosophical Transactions* 14 (20 August 1684):672-76.

________. "A Letter from Mr. St George Ash, Sec. of the Dublin Society, to one of the Secretaries of the Royal Society; concerning a Girl in Ireland, who has several Horns growing on her Body." *Philosophical Transactions* 15 (26 November 1685):1202-04.

________. "Part of a Letter from Dr. Ashe, Lord Bishop of Cloyne, dated March the 26th 1687. concerning the Effects of Imagination, the Vertues of Mackenboy, &c." *Philosophical Transactions* 20 (August 1698):293-94.

Ashe, St. George, et al. "Observations of the Solar Eclipse July the 2. 1684. at Oxford, sent in a Letter from Dr. Edw. Bernard, Astron. Prof. Oxon. to Mr. John Flamsteed, Astron. Reg. At Lisbon by Mr. Jacobs; at Dublin, by Mr. Ash and Mr. Molyneaux; and at Tredagh by Mr. Osborne." *Philosophical Transactions* 14 (20 October 1684):747-49.

Aubrey, John. *"Brief Lives," chiefly of Contemporaries, set down by John Aubrey, between the Years 1669 & 1696*. Edited by Andrew Clark. 2 vols. Oxford: Clarendon Press, 1898.

Baker, Tho[mas]. *The Geometrical Key: or the Gate of Equations Unlock'd: A New discovery of the Construction of all Equations, howsoever affected, not exceeding the fourth Degree; viz. of, Linears, Quadratics, Cubics, Biquadratics; And the finding of all their Roots, as well false, as true; without the use of Mesolabe, Trisection of Angles; without Reduction, Depression, or any other previous preparation of Equations, by a Circle, and any (and that but one only) Parabole*. London: Printed by J. Playford, for R. Clavel, at the Peacock in S. Paul's Church-yard, 1684.

Barrow, Isaac. *A Brief Exposition on the Creed, the Lord's Prayer, and Ten Commandments. To which is added The Doctrine of the Sacraments*. London: Printed by J. H. for Brabazon Aylmer, at the Three Pigeons, over against the Royal Exchange in Cornhill, 1697.

________. *Lectiones XVIII, Cantabrigiae in Scholis publicis habitae; in quibus Opticorum Phaenomenon Genuinae Rationes investigantur, ac exponuntur. Annexae sunt Lectiones aliquot Geometricae*. London: Typis Gulielmi Godbid, & prostant venales apud Johannem Dunmore, & Octavianum Pulleyn Juniorem, 1669.

________. *Lectiones Geometricae: In quibus (praesertim) Generalia Curvarum Linearum Symptomata Declarantur*. London: Typis Gulielmi Godbid, & prostant venales apud Johannem Dunmore, 1670.

________. *Lectiones Opticae & Geometricae: In quibus Phaenomenon Opticorum Genuinae Rationes investigantur, ac exponuntur: Et Generalia Curvarum Linearum Symptomata declarantur*. London: Typis Guilielmi Godbid, & prostant venales apud Robertum Scott, in vico Little Britain, 1674.

________. *The Theological Works of Isaac Barrow, D.D., Master of Trinity College, Cambridge*. Edited by Alexander Napier. 9 vols. Vol. 1: *Fifteen Sermons on Several Occasions*. Vol. 5: *Seventeen Sermons on the Creed*. Vol. 7: *An Exposition of the Creed; The Lord's Prayer; The Decalogue; The Doctrine of the Sacraments*. Cambridge: Cambridge University Press, 1859.

________. *The Usefulness of Mathematical Learning explained and demonstrated: Being Mathematical Lectures Read in the Publick Schools at the University of Cambridge... To which is prefixed, The Oratorical Preface of Our Learned Author, spoke before the University on his being elected Lucasian Professor of the Mathematics*. Translated by John Kirkby. London: Printed for Stephen Austen, at the Angel and Bible in St. Paul's Church-yard, 1734; facsimile reprint ed., London: Frank Cass and Company Limited, 1970.

________, ed. *Archimedis Opera: Apollonii Pergaei Conicorum Libri IIII. Theodosii Sphaerica: Methodo Nova Illustrata, & Succincte Demonstrata*. London: Excudebat Guil. Godbid, voeneunt apud Rob. Scott, in vico Little Britain, 1675.

________, ed. *Euclidis Elementorum Libri XV. breviter demonstrati,... Et prioribus mendis typographicis nunc demum purgati*. London: Typis J. Redmayne: Prostant autem apud J. Williams ad Insigne Coronae in Coemeterio D. Pauli, & J. Dunmore ad Insigne Trium Bibliorum in vico vulgo vocato Ludgate-street, 1678.

B[eale], J[ohn]. *Herefordshire Orchards, A Patterne For all England. Written in an Epistolary Address to Samuel Hartlib Esq.* London: Printed by Roger Daniel, 1657.

Birch, Thomas. *The History of the Royal Society of London for Improving of Natural Knowledge From its First Rise. In which The most considerable of those Papers communicated to the Society, which have hitherto not been published, are inserted in their proper order, as a supplement to The Philosophical Transactions.* 4 vols. London: n.p., 1756-57; facsimile reprint ed., with an introduction by A. Rupert Hall and additional bibliographical note by Marie Boas Hall, The Sources of Science, no. 44, New York: Johnson Reprint Co., 1968.

B[oyle], R[obert]. *A Free Enquiry Into the Vulgarly Receiv'd Notion of Nature Made in an Essay, Address'd to a Friend.* London: Printed by H. Clark, for John Taylor at the Globe in St. Paul's Church-yard, 1685/86.

________. *Reflections upon a Theological Distinction. According to which it is said, that some Articles of Faith are above Reason, but not against Reason. In a Letter to a Friend. In The Works of the Honourable Robert Boyle... To which is prefixed the Life of the Author*, 5:541-49. 6 vols. London: Printed for W. Johnston, S. Crowder, T. Payne, G. Kearsley, J. Robson, B. White, T. Becket and P. A. Hondt, T. Davies, T. Cadell, Robinson and Roberts, Richardson and Richardson, J. Knox, W. Woodfall, J. Johnson, and T. Evans, 1772.

[________]. *A Discourse of Things above Reason Inquiring Whether a Philosopher should admit there are any such. To which are annexed by the Publisher (for the Affinity of the Subjects) Some Advices About judging of Things said to Transcend Reason.* London: Printed by E. T. and R. H. for Jonathan Robinson at the Golden Lion in S. Paul's Church-Yard, 1681.

________. *The Excellency of Theology, Compar'd with Natural Philosophy (as both are Objects of Men's Study.) Discours'd of In a Letter to a Friend. To which are annex'd Some Occasional Thoughts about the Excellency and Grounds of the Mechanical Hypothesis.* London: Printed by T. N. for Henry Herringman, at the Anchor in the Lower Walk of the New Exchange, 1674.

[________]. *Some Considerations about The Reconcileableness of Reason and Religion To Which is annex'd by the Publisher, A Discourse of Mr. Boyle about The Possibility of the Resurrection.* London: Printed by T. N. for H. Herringman, at the Anchor in the Lower Walk of the New Exchange, 1675.

________. *Tracts written by the Honourable Robert Boyle, containing New Experiments, touching the relation betwixt flame and air. And about explosions. An hydrostatical discourse occasion'd by some objections of Dr. Henry More against some explications of new experiments made by the author of these tracts: To which is annex't, An hydrostatical letter dilucidating an experiment about the way of weighing water in water. New experiments, Of the positive or relative levity of bodies under water. Of the air's spring on bodies under water. About the differing pressure of heavy solids and fluids.* London: Printed for Richard Davis, bookseller in Oxon, 1672.

________. *The Works of the Honourable Robert Boyle.... To which is prefixed the Life of the Author.* 6 vols. London: Printed for W. Johnston, S. Crowder, T. Payne, G. Kearsley, J. Robson, B. White, T. Becket and P. A. Hondt, T. Davies, T. Cadell, Robinson and Roberts, Richardson and Richardson, J. Knox, W. Woodfall, J. Johnson, and T. Evans, 1772.

Burnet, Gilbert. *Bishop Burnet's History of His Own Time.* [Edited by Sir Thomas Burnet]. 8 vols. Edinburgh: Printed by Hamilton, Balfour, and Neill, 1753.

Chillingworth, William. *The Religion of Protestants A Safe way to Salvation. Or, An Answer to a Book Entituled Mercy and Truth, or, Charity maintain'd by Catholiques: Which pretends to prove the Contrary. To which is Added The Apostolical Institution of Episcopacy. As also IX. Sermons, The first Preached before His Majesty King Charles the First, the other Eight upon special and eminent Occasions.* 4th ed. London: Printed by Andrew Clark, for Richard Chiswell at the Rose and Crown in St. Paul's Church-yard, 1674.

Evelyn, John. *The Diary of John Evelyn: Now first printed in full from the manuscripts belonging to Mr. John Evelyn....* Edited by E. S. De Beer. 6 vols. Oxford: Clarendon Press, 1955.

Firth, C. H. and Rait, R. S., comp. and ed. *Acts and Ordinances of the Interregnum, 1642-1660.* 3 vols. London: H. M. Stationery Office, 1911.

Flamsteed, John. "An Accompt of such of the more notable Celestial Appearances of the Year 1670, as will be conspicuous in the English horizon; written by the Learned and Industrious Mr. John Flamstead Novemb. 4 1669. and by him addressed and recommended for encouragement, to the Right Honorable, the Lord Viscount Brouncker, as President of the Royal Society." *Philosophical Transactions* 4 (17 January 1670):1099-1112.

________. *Atlas Coelestis.* London: n.p., 1729.

________. "De Lunae ad Pleiades reliquas Applicationibus, Anno 1672 observandis, promissa appendicula Johannis Flamsteed." *Philosophical Transactions* 6 (22 January 1671/72):3061-63.

[________]. *The Doctrine of the Sphere, Grounded on the Motion of the Earth, And the Antient Pythagorean or Copernican System of the World. In Two Parts.* London: Printed by A. Godbid and J. Playford, 1680. In Sir Jonas Moore, *A New Systeme of the Mathematicks: Containing I. Arithmetick, as well Natural and Decimal, as in Species, or the Principles of Algebra. II. Practical Geometry together with the first Six Books of Euclid's Elements, as also the Eleventh and Twelfth, symbolically demonstrated. III. Trigonometry Plain and Spherical. IV. Cosmography, or a Description of the Heavens. V. Navigation, or Sailing by a Plain or Mercator's Chart; as also by the Arch of a Great Circle, &c. VI. The Doctrine of the Sphere, grounded on the Motion of the Earth, according to the Old Pythagorean and Copernican Systeme. VII. Astronomical Tables, with Tables of Logarithms, Natural and Artificial Sines and Tangents, and Versed Sines. VIII. A New Geography, or a Description of the most Eminent Countries and Coasts of the World, with Maps of them, and Tables of their Latitude and Longitude.* London: Printed by A. Godbid and J. Playford for Robert Scott, Bookseller in Little Britain, 1681.

________. "An Extract of Mr. Flamstead's Letter written to the Publisher from Derby Novemb 16. 1672. concerning the Appulses, by him calculated for A. 1673. of the Moon, and the other Planets to Fixed Stars; together with an Observation of the Planet Mars, made by the same." *Philosophical Transactions* 7 (16 December 1672):5118-24.

________. *Historia Coelestis Britannica.* 3 vols. London: Typis H. Meere, 1725.

________. "Lunae ad Fixas Appulsus Visibiles, nec non arctiores juxta eas Transitus, observabiles A. 1672: Praedicti, & ad Meridianum Latitudinemque Londini, e Tabulis Carolinis accurate supputati a Joh. Flamstedio Derbiensi-Anglo." *Philosophical Transactions* 6 (20 November 1671):2297-3001.

________. "Lunae ad Fixas Saturnumque, Anno 1671. Appulses observabiles, praedicti, & ad Meridianum Latitudinemque Londini accurate supputati a Joh. Flamsteadio Derbiensi-Anglo." *Philosophical Transactions* 5 (12 December 1670):2029-34.

________. "Mr. Flamstead's account of his Observations of the late Comet, sent in a letter to the Publisher, Greenwich, May 18. 1677." *Philosophical Transactions* 12 (26 May 1677):873-75.

Glanvill, Joseph. *Essays on several Important Subjects in Philosophy and Religion.* London: Printed by J. D. for John Baker, at the Three Pidgeons, and Henry Mortlock, at the Phoenix in St. Pauls Church-Yard, 1676; facsimile reprint ed., with a new introduction by Richard H. Popkin, New York: Johnson Reprint Corporation, 1970.

[________]. *Logou Threskeia Or, A Seasonable Recommendation, and Defence of Reason, In the Affairs of Religion; Against Infidelity, Scepticism, and Fanaticisms of all sorts.* London: Printed by J. M. for James Collins at the Kings-bead in Westminster-Hall, 1670; facsimile reprint ed., Hildesheim: Georg Olms Verlag, 1970.

________. *Philosophia Pia; or a Discourse of the Religious Temper, and Tendencies of the Experimental Philosophy which is Profest by the Royal Society. To which is annext A Recommendation and Defence of Reason in the Affairs of Religion.* London: Printed by J. Macock for James Collins at the Kings-Arms in Ludgate street near the West end of S. Pauls, and at his Shop at the Kings Head in Westminster-Hall, 1671; facsimile reprint ed., Hildesheim: Georg Olms Verlag, 1970.

________. *Plus Ultra: Or The Progress and Advancement of Knowledge Since the Days of Aristotle. In an Account of some of the most Remarkable Late Improvements Of Practical, Useful Learning: To Encourage Philosophical Endeavours. Occasioned By a Conference with one of the Notional Way.* London: Printed for James Collins at the Kings-Head in Westminster-Hall, 1668; facsimile reprint ed., edited by Jackson I. Cope, Gainesville, FL: Scholars' Facsimiles & Reprints, 1958.

________. *A Praefatory Answer to Mr. Henry Stubbe, The Doctor of Warwick. Wherein the Malignity of his Temper, the Hypocrisie of his Pretences, the Falshood of his Reports, And the Impertinency of his Arguings & Quotations in his Animadversions on Plus Ultra, Are discovered.* London: Printed by A. Clark, for J. Collins, at the Kings Arms in Ludgate-street, near the West-end of S. Pauls; and at the Kings Head in Westminster-hall, 1671.

________. *Seasonable Reflections And Discourses In Order to the Conviction, & Cure Of the Scoffing, & Infidelity Of a Degenerate Age.* London: Printed by R. W. for H. Mortlock at the Phoenix in St. Paul's Church-yard, and the White-Hart in Westminster-Hall, 1676.

Gunther, Robert W. T., ed. *Early Science in Oxford.* Vol. 12: *Dr. Plot and the Correspondence of the Philosophical Society of Oxford.* Oxford: Printed for the subscribers, 1939.

________, ed. "Transactions of the Philosophical Society of Oxford, 26 October 1683 to 14 April 1685 and 23 April 1685 to 3 June 1690." In *Early Science in Oxford*. Vol. 4: *The Philosophical Society*, pp. 17-220. Oxford: Printed for the subscribers, 1925.

Hearne, Thomas. *Reliquiae Hearnianae: The Remains of Thomas Hearne, M.A., of Edmund Hall, Being Extracts from his MS. Diaries....* Compiled by Philip Bliss. 2 vols. Oxford: Printed for the Editor, By James Wright, Printer to the University, 1857.

Hooker, Richard. *The Folger Library Edition of The Works of Richard Hooker*. W. Speed Hill, gen. ed. 4 vols. Cambridge: Harvard University Press, The Belknap Press, 1977-82. Vol. 1: *Of the Laws of Ecclesiastical Polity: Preface, Books I-IV*. Edited by Georges Edelen.

Horrocks, Jeremiah. *Jeremiae Horroccii, Liverpoliensis Angli, ex Platinatu Lancastriae Opera Posthuma; viz. Astronomia Kepleriana, defensa & promota. Excerpta ex Epistolis ad Crabtraeum suum. Observationum Coelestium Catalogus. Lunae Theoria nova. Accedunt Guilielmi Crabtraei, Mancestriensis, Observationes Coelestes. In calce adjiciuntur Johannis Flamstedii, Derbiensis, De Temporis Aequatione Diatriba. Numeri ad Lunae Theoriam Horroccianam.* Edited by John Wallis. London: Typis Gulielmi Godbid, Impensis J. Martyn Regalis Societatis Typographi, ad insigne Campanae in Coemeterio D. Pauli, 1673.

Jackson, Thomas. *The Eternal Truth of Scriptures, and Christian Beleefe, Thereon Wholly Depending, Manifested By It Owne Light. Delivered in Two Bookes Of Commentaries upon the Apostles Creede; The former, Containing the positive grounds of Christian Religion in generall, cleared from all exceptions of Atheists or Infidels. The later, Manifesting the Grounds of Reformed Religion to be so firme and sure, that the Romanist cannot oppugne them, but with the utter overthrow of the Romish Church, Religion, and Faith.* London: Printed by W. Stansby and are to bee sold by John Budge, at the Great South doore of Pauls, and at Britaines-Bursse, 1613.

________. *The Works of the Reverend and Learned Divine, Thomas Jackson, D.D. Sometime President of Corpus Christi College in Oxon; Such as were, and such as Never before were Printed. . . With the Authors Life, And A Large and Useful Table to the Whole*, vol. 1: *Justifying Faith: or, The Faith by which the just do live. A Treatise Containing a Description of the Nature, Properties & Conditions of Christian Faith. With A Discovery of mis-perswasions, breeding Presumption or Hypocrisie; and means how Faith may be planted in Unbelievers.* London: Printed by Andrew Clark, for John Martyn, Richard Chiswell, and Joseph Clark; and are to be sold at the Bell, and at the Rose and Crown in St. Paul's Church-yard, and at the Star in Little Britain, 1673.

Laney, Benjamin. *Five Sermons, Preached before His Majesty at Whitehall. Published Severally by Command. And Now Printed together, tending all to give satisfaction in Certain Points to such who have thereupon endeavoured to unsettle the State and Government of the CHURCH.* London: Printed for Timothy Garthwait, 1669.

________. *A Sermon Preached before His Majesty at Whitehall March 9th. 1661. . .Published by His Majesties Special Command.* London: Printed by R.N. for Timothy Garthwait, at the Kings-Head in St. Pauls Churchyard, 1662.

________. *A Sermon Preached at Whitehall before His Majesty March 27th 1664.* London: Printed for Timothy Garthwait, at Kings-Head in St. Paul's Church-yard, 1665.

________. *A Sermon Preached before the King at White-Hall. Being the last Sermon preached at Court ....* London: Printed for Henry Brome at the Gun at the West end of S. Pauls, 1675.

Locke, John. *Of the Conduct of the Understanding*. Edited by Francis W. Garforth. New York: Columbia University Teachers College Press, no. 31, 1966.

________. *An Early Draft of Locke's Essay: Together with Excerpts from His Journals*. Edited by R. I. Aaron and Jocelyn Gibb. Oxford: Clarendon Press, 1936.

________. *The Educational Writings of John Locke*. Edited by James L. Axtell. Cambridge: Cambridge University Press, 1968.

________. *An Essay Concerning Human Understanding*. Edited by John W. Yolton. 2 vols. New York: Dutton, Everyman's Library, 1971.

________. *An Essay Concerning the Understanding, Knowledge, Opinion, and Assent*. Edited with an introduction by Benjamin Rand. Cambridge: Harvard University Press, 1931.

More, Henry. *An Antidote Against Atheisme, or An Appeal to the Natural Faculties of the Minde of Man, where there be not a God*. London: Printed by Roger Daniel, at Lovell's Inne in Pater-noster-Row, 1653.

Newton, Isaac. *The Correspondence of Isaac Newton*. Edited by H. W. Turnbull. 4 vols. Cambridge: Published for the Royal Society at the University Press, 1959.

Norris, John. *An Account of Reason and Faith: In Relation to the Mysteries of Christianity*. London: Printed for S. Manship, at the Ship near the Royal Exchange in Cornhill, 1697.

Oldenburg, Henry. *The Correspondence of Henry Oldenburg*. Translated and edited by A. Rupert Hall and Marie Boas Hall. 13 vols. Vols. 1-9, Madison: The University of Wisconsin Press, 1965-75. Vols. 10-11, London: Mansell, 1975-77. Vol. 13, London: Taylor & Francis, 1986.

[Payne, William]. *A Discourse Concerning the Adoration of the Host As it is Taught and Practiced in the Church of Rome. Wherein an Answer is given T. G. on that Subject, and to Monsieur Boileau's late book De Adoratione Eucharistiae. Paris 1685*. London: Printed for Brabazon Aylmer at the Three Pigeons, against the Royal Exchange in Cornhill, 1685.

[________]. *A Discourse of the Communion in One Kind: In Answer to a Treatise of the Bishop of Meaux's, of Communion under both Species Lately Translated into English*. London: Printed for Brabazon Aylmer, at the three Pidgeons over against the Royal Exchange in Cornhil, 1687.

[________]. *A Discourse of the Sacrifice of the Mass*. London: Printed for Brabazon Aylmer, at the Three Pidgeons over against the Royal Exchange in Cornhil, 1688.

________. *Learning and Knowledge Recommended to the Scholars of Brentwood School in Essex; in a Sermon Preached at their First Feast, June 29, 1682*. London: Printed for Walter Kettilby at the Bishop's Head in St. Paul's Church-yard, 1682.

________. *The Mystery of the Christian Faith and of the Blessed Trinity Vindicated and The Divinity of Christ Proved, In Three Sermons, Preach'd at Westminster-Abbey upon Trinity Sunday, June the 7th and September 21, 1696. By the late Reverend William Payne, D.D. In the Press before his Death, and by himself ordered to be Published.* London: Printed for Richard Cumberland, at the Angel in St. Pauls Church-Yard, 1697.

________. *A Practical Discourse of Repentance, Rectifying the Mistakes about it, especially such as lead either to Despair or Presumption. Perswading and Directing to the True Practice of it, and Demonstrating the Invalidity of a Death-Bed Repentance.* London: Printed for Samuel Smith at the Princes Arms in St. Pauls Church-Yard, 1693.

Pearson, John. *An Exposition of the Creed.* 5th ed. London: Printed by J. M. for John Williams, at the Crown in St. Paul's Church-yard, 1683.

________. *An Exposition of the Creed.* 6th ed. London: Printed by J. M. for J. Williams, and are to be Sold by Tho. Sawbridge at the Three Flower-de-Luces in Little Britain, 1692.

________. *Opera Posthuma Chronologica, &c. viz. De Serie et Successione Primorum Romae Episcoporum Dissertationes Duae: Quibus praefiguntur Annales Paulini, et Lectiones in Acta Apostolorum. Singula Praelo tradidit, edenda curavit & Dissertationis novis Additionibus auxit H. Dodwellus, A.M. Dubliniensis: Cujus etiam accessit De eadem Successione usque ad Annales Cl. Cestriensis Cyprianicos Dissertatio Singularis.* London: Typis S. Roycroft, L. L. Oriental. Typographi Regis; Impensis R. Clavell, & B. Tooke, in Coemeterio D. Pauli, 1688.

________. *Vindiciae Epistolarum S. Ignatii...Accesserunt Isaaci Vossii Epistolae Duae Adversus David Blondellum.* Cambridge: Typis Joann Hayes, Prostant Londini, apud Guil. Wells & Rob. Scott, ad Insignia Principis in vico Little Britain dicto, 1672.

Pell, John. *Controversiae de vera Circuli mensura Anno [1644] exortae, inter Christianum Severini Longomontanum, Cimbrum, Superiorum Mathematum in Regia Danorum Academia Hauniensi, Professoren publicum et Ioannem Pellium, Coritano-Regnum, Anglum, Matheseos, in Illustri Amstelodamensium Gymnasio, Professorem publicum. Pars Prima.* Amsterdam: Apud Ioannem Blaeu, 1647.

________. "An Idea of Mathematicks." In John Dury, *The Reformed Librarie Keeper With a Supplement to the Reformed-School, As subordinate to Colleges in Universities. Whereunto is added I. An Idea of Mathematicks. II. The description of one of the chiefest Libraries which is in Germanie, erected and ordered by one of the most Learned Princes in Europe*, pp. 33-46. London: Printed by William Du-Gard, and are to bee sold by Rob. Littleberrie at the sign of the Unicorn in Little Britain, 1650.

Pepys, Samuel. *The Diary of Samuel Pepys.* Edited by Robert Latham and William Matthews. 9 vols. Berkeley: University of California Press, 1970-76.

Pierce, Thomas. *A Decad of Caveats to the People of England, of General Use in All Times, but most seasonable in These. As having {A Tendency to the satisfying such as are not content with the Present Government, as it is by Law established. {An Aptitude to the Setling the Minds of such as are but Seekers and Erraticks in Religion. {An Aim at the Uniting of our Protestant-Dissenters in Church and State. Whereby The worst of All Conspiracies lately rais'd against Both, may be made the greatest Blessing, which could have happen'd to either of them. To which is added an Appendix In order to the Conviction of those Three Enemies to the Deity: The Atheist, The Infidel, and The*

Setter up of Science to the Prejudice of Religion. London: Printed for Richard Davis Bookseller in Oxford, 1679.

Pigot, Tho[mas]. "An Account of the Earthquake that happened at Oxford and the parts adjacent Sept. 17. 1683. by a fellow of a College in that University, and of the Royal Society." *Philosophical Transactions* 13 (20 September 1683):311-21.

Pope, Walter. *The Life of the Right Reverend Father in God Seth, Lord Bishop of Salisbury, and Chancellor of the Most Noble Order of the Garter. With a brief Account of Bishop Wilkins, Mr. Lawrence Rooke, Dr. Isaac Barrow, Dr. Tubervile, And others.* London: Printed for William Keblewhite, at the Swan in St. Paul's Church-yard, 1697; facsimile reprint ed., edited by J. B. Bamborough, Oxford: Basil Blackwell for the Luttrell Society, 1961.

[Puccini, Vincent]. *The Life of St. Mary Magdalene of Pazzi, a Carmelite Nunn, Newly translated out of Italian by the Reverend Father Lezin de Sainte Scholastique, Provincial of the Reformed Carmelites of Touraine. At Paris For Sebat. Cramoisy in St. James Street, at the Sign of Fame. 1670. And now done out of French: With a preface concerning the nature, causes, concomitants, and consequences of Ecstasy and Rapture, and a brief discourse added about discerning and trying the spirits, whether they be of God.* [Translated by Thomas Smith]. London: Printed, And are to be sold by Randal Taylor near Stationers Hall, 1687.

[Rahn, Johann Heinrich]. *An Introduction to Algebra Translated out of the High-Dutch into English,... Also A Table of Odd Numbers less than One Hundred Thousand, Shewing Those are Incomposit, and Resolving the rest into their Factors or Coefficients, &c.* Translated by Tho[mas] Brancker. Edited by D[r]. P[ell]. London: Printed by W. G. for Moses Pitt at the White-Hart in Little Britain, 1668.

[Ray, John]. *Catalogus Plantarum Circa Cantabrigiam nascentium: In quo exhibentur Quotquot hactenus inventae sunt, quae vel sponte proveniunt, vel in agris feruntur; Una cum Synonymis selectioribus, locis natalibus & observationibus quibusdam oppido raris. Adjiciuntur in gratiam tyronum, Index Anglico-latinus, Index locorum, Etymologia nominum, & Explicatio quorundam terminorum.* Cambridge: Excudebat Joann. Field, celeberrimae Academiae Typographus. Impensis Gulielmi Nealand, Bibliopolae, 1660.

________. *A Collection of English Proverbs Digested into a convenient Method for the speedy finding any one upon occasion; with Short Annotations. Whereunto are added Local Proverbs with their Explications, Old Proverbial Rhythmes, Less known or Exotick Proverbial Sentences, and Scottish Proverbs.* Cambridge: Printed for John Hayes, Printer to the University, for W. Morden, 1670.

________. *A Collection of English Words, Not Generally used, with their Significations and Original, in two Alphabetical Catalogues, The one of such as are proper to the Northern, the other to the Southern Counties. With Catalogues of English Birds and Fishes: And an Account of the preparing and refining such Metals and Minerals as are gotten in England.* London: Printed by H. Bruges. For Tho. Burrell at the Golden-Ball under St. Dunstans Church in Fleetstrete, 1674.

________. *The Correspondence of John Ray: Consisting of Selections from the Philosophical Letters Published by Dr. Derham, and Original Letters of John Ray in the Collection of the British Museum.* Edited by Edwin Lankester. London: Printed for the Ray Society, 1948.

________. *Further Correspondence of John Ray*. Edited by Robert W. T. Gunther. London: Printed for the Ray Society, 1928.

________. *Historia Plantarum Species hactenus editas aliasque insuper multas noviter inventas & descriptas complectens. In qua agitur primo De Plantis in genere, Earumque Partibus, Accidentibus & Differentiis; Deinde Genera omnia tum summa tum subalterna ad Species usque infimas, Notis suis certis & Characteristicis Definita, Methodo Naturae vestigiis insistente disponuntur; Species singulae accurate describuntur, obscura illustrantur, omissa supplentur, superflua resecantur, Synonyma necessaria adjiciuntur; Vires denique & Usus recepti compendio traduntur*. Vol. 1. London: Typis Mariae Clark: Prostant apud Henricum Faithorne Regiae Societatis Typographum ad Insigne Rosae in Coemeterio D. Pauli, 1686.

________. *Historiae Plantarum Tomus Secundus: Cum Duplici Indice Generali Altero Nominum & Synonymorum praecipuorum; Altero Affectuum & Remediorum: Accessit Nomenclator Botanicus Anglo-Latinus*. Vol. 2. London: Typis Mariae Clark: Prostant apud Henricum Faithorne Regiae Societatis Typographum, ad insigne Rosae in Coemeterio D. Pauli, 1688.

________. *Historiae Plantarum Tomus Tertius: qui est Supplementum Duorum praecedentium: Species omnes vel omissas, vel post Volumina illa evulgata editas, praeter inumeras fere novas & indictas ab Amicis communicatas complectens: cum Synonymis necessariis, et Usibus in Cibo, Medicina, & Mechanicis: Additio ad Opus consummandum Generum Indice copioso. Accessit Historia Stirpium Ins. Luzonis & reliquarum Philippinarum a R. P. Geo. Jos. Camello, Moravo-Brunensi, S.J. conscripta. Item D. Jos. Pitton Tournefort, M.D. Parisiensis, & in Horto Reg. Botanices Professoris, Corollarium Institutionum Rei Herbariae*. Vol. 3. London: Apud Sam. Smith & Benj. Walford, Reg. Soc. Typographos, ad Insignia Principis in Area Boreali D. Pauli, 1704.

________. *Methodus Plantarum Nova, Brevitatis & Perspicuitatis Causa Synoptice in Tabulis Exhibita; Cum notis Generum tum summorum tum subalternorum Characteristicis, Observationibus nonnullis de seminibus Plantarum & Indice Copioso*. London: Impensis Henrici Faithorne, & Joannis Kersey ad insigne Rosae in Coemeterio D. Pauli, 1682.

________. *Miscellaneous Discourses Concerning the Dissolution and Changes of the World Wherein The Primitive Chaos and Creation, the General Deluge, Fountains, Formed Stones, Sea-Shells found in the Earth, Subterraneous Trees, Mountains, Earthquakes, Vulcanoes, the Universal Conflagration and Future State, are largely Discussed and Examined*. London: Printed for Samuel Smith, at the Prince's Arms in St. Paul's Church Yard, 1692.

________. *Three Physico-Theological Discourses, Concerning I. The Primitive Chaos, and Creation of the World. II. The General Deluge, its Causes and Effects. III. The Dissolution of the World, and Future Conflagration. Wherein Are largely Discussed the Production and Use of Mountains; the Original of Fountains, of Formed Stones, and Sea-Fishes, Bones and Shells found in the Earth; the Effects of particular Floods and Inundations of the Sea; the Eruptions of Vulcano's; the Nature and Causes of Earthquakes: With an Historical Account of those Two late Remarkable Ones in Jamaica and England. With Practical Inferences*. 2nd. ed. London: Printed for Sam. Smith, at the Princes Arms in St. Paul's Church-yard, 1693.

________. *The Wisdom of God Manifested in the Works of the Creation. Being the Substance of some common Places delivered in the Chappel of Trinity-College, in Cambridge.* London: Printed for Samuel Smith, at the Princes Arms in S. Pauls Church-Yard, 1691.

Rigaud, Stephen Peter, ed. *Correspondence of Scientific Men of the Seventeenth Century, Including Letters of Barrow, Flamsteed, Wallis, and Newton, Printed From the Originals in the Collection of the Right Honourable the Earl of Macclesfield.* 2 vols. Oxford: Oxford University Press, 1841.

Smith, Edward. "A Relation of an extraordinary effect of the power of Imagination: Communicated by Mr. Edward Smith, Secretary to the Philosophical Society at Dublin, as it was brought before that Company, by Mr. St. George Ash. R. Soc. S. who had seen the thing." *Philosophical Transactions* 16 (April-June 1687):334.

Smith, Tho[mas]. "An Account of the City of Prusa in Bithynia, and a continuation of the Historical Observations relating to Constantinople." *Philosophical Transactions* 14 (20 January 1683/84):431-54.

________. *An Account of the Greek Church, as to Its Doctrine and Rites of Worship: With Several Historicall Remarks interspersed, relating thereunto. To which is added An Account of the State of the Greek Church, under Cyrillus Lucaris Patriarch of Constantinople, with a Relation of his Sufferings and Death.* London: Printed by Miles Flesher for Richard Davis in Oxford, 1680.

________. *Admodum Reverendi & Doctissimi Viri, D. Roberti Huntingtoni, Epistolae: et Veterum Mathematicorum, Graecorum, Latinorum, & Arabum, Synopsis: Collectore Viro Clarissimo & Doctissimo, D. Edwardo Bernardo, Astronomiae in Academia Oxoniensi Professore Saviliano. Praemittuntur D. Huntingtoni & D. Bernardi Vitae.* London: Typis Gul. Bowyer, Impensis A. & J. Churchill, ad Insigne Nigri Cygni in Vico vulgò vocato Paternoster-Row, 1704.

________. "A conjecture about an under-Current at the Streights-Mouth read before the Oxford Society, Dec. 21. 1683." *Philosophical Transactions* 14 (20 April 1684):564-66.

________. "Historical Observations relating to Constantinople." *Philosophical Transactions* 13 (20 October 1683):335-46.

________. "A Journal of a Voyage from England to Constantinople, made in the Year 1668 by T. Smith D.D. & F.R.S." *Philosophical Transactions* 19 (July 1697):597-619.

________. *Remarks Upon the Manners, Religion and Government Of the Turks. Together with A Survey of the Seven Churches of Asia, As they now lye in their Ruines: and A Brief Description of Constantinople.* London: Printed for Moses Pitt, at the Angel in St. Pauls Church-yard, 1678.

________. *A Sermon Concerning the Doctrine, Unity, and Profession of the Christian Faith. Preached before the University of Oxford, To which is added an Appendix concerning the Apostles Creed.* London: Printed for Walter Kettilby, at the Bishop's Head in St. Paul's Church-Yard, 1682.

________. *A Sermon of the Credibility of the Mysteries of the Christian Religion. Preached Before a Learned Audience.* London: Printed by Tho. Roycroft for Ric. Davis Bookseller in Oxford, 1675.

Sprat, Thomas. *A Discourse Made by the Ld Bishop of Rochester to the Clergy of his Diocese, at His Visitation in the Year 1695. Published at their Request.* London: In the Savoy: Printed by Edw. Jones, 1696.

________. *History of the Royal-Society of London, For the Improving of Natural Knowledge.* London: Printed by T.R. for J. Martyn at the Bell without Temple-bar, and J. Allestry at the Rose and Crown in Duck-Lane, Printers to the Royal Society, 1667; facsimile reprint ed., edited by Jackson I. Cope and Harold Whitmore Jones, St. Louis: Washington University Studies, 1958.

________. *A Sermon Preach'd to the Natives of The County of Dorset, Residing in and about the Cities of London and Westminster, At St. Mary Le Bowe, on Dec. 8, 1692. Being the Day of their Anniversary Feast.* London: In the Savoy; Printed by Edward Jones, and sold by him and Randal Taylor near Stationers-Hall, 1693.

Tillotson, John. *The Rule of Faith: or an Answer to the Treatise of Mr. I. S. Entituled, Sure-footing, &c....To which is Adjoined A Reply to Mr. I. S. his 3d Appendix, &c. By Edward Stillingfleet B.D. one of His Majesties Chaplains in Ordinary.* London: Printed by A. Maxwell for Sa. Gellibrand, at the Golden Ball in St. Pauls Church-yard, 1666.

________. *A Seasonable Vindication of the B. Trinity. Being an Answer to this Question, "Why do you believe the Doctrine of the Trinity?" Collected from the Works of the Most Reverend, Dr.* John Tillotson, *Late Lord Archbishop of Canterbury. And the Right Reverend, Dr.* Edward Stillingfleet *Now Lord Bishop of Worcester.* London: Printed for B. Aylmer, at the Three Pigeons against the Royal Exchange in Cornhill, 1697.

________. *Sermons Preach'd Upon Several Occasions.* London: Printed by A. M. for Sa: Gellibrand, at the Golden Ball in St. Paul's Church-yard, 1671.

________. "Sermons 218-223." In *The Works of Dr. John Tillotson, Late Archbishop of Canterbury*, 9:173-279. Edited by Thomas Birch. 10 vols. London: Printed by J.F. Dove, St. John's Square for Richard Priestley, High Holborn, 1820.

[Tindal, Matthew]. *A Letter to the Reverend the Clergy of both Universities, Concerning the Trinity and the Athanasian Creed. With Reflections on all the late Hypotheses, particularly Dr. W's, Dr. S--th's; The Trinity placed in its due Light; The 28 Propositions; The calm Discourse of a Trinity in the Godhead, and the Defence of Dr. Sherlock's Notions. With a short Discourse Concerning Mysteries.* London: n. p., 1694.

[Toland, John]. *Christianity not Mysterious: or a Treatise Shewing, That there is nothing in the Gospel Contrary to Reason, Not Above it: And that no Christian Doctrine can be properly call'd A Mystery.* London: n. p., 1696; facsimile reprint ed., edited by Rene Wellek, British Philosophers and Theologians of the 17th & 18th Centuries, New York: Garland Publishing, Inc., 1978.

Vossius, Isaac. *A Treatise Concerning the Motion of the Seas and Winds.* Translated by Arch. Lovel. London: Printed for Henry Brome, at the Gun near the West end of St. Pauls-Church, 1677.

Wallis, John. *Arithmetica Infinitorum, Sive Nova Methodus Inquirendi in Curvilineorum Quadraturam, aliaq; difficiliora Matheseos Problemata.* Oxford: Typis Leon: Lichfield Academiae Typographi, Impensis Tho. Robinson, 1656. In John Wallis, *Operum*

Mathematicorum Pars Altera: Qua Continentur {De Angulo Contactus & Semicirculi, Disquisitio Geometrica. {De Sectionibus Conicis Tractatus. {Arithmetica Infinitorum: sive de Curvilineorum quadratura, &c. {Ecclipseos Solaris Observatio. Oxford: Typis Leon Lichfield Academiae Typographi, Impensis Tho. Robinson, 1656.

________. *A Brief and Easie Explanation Of The Shorter Catechism, Presented By the Assembly of Divines at Westminster, to both Houses of Parliament, And By them Approved. Wherein the meanest Capacities may in a speedy and easie way be brought to understand the Principles of Religion. In imitation of a Catechisme, formerly Published by Mr. Herbert Palmer, B.D. and late Master of Queens Colledge.* 8th ed. London: Printed for Peter Parker, at the entrance into Popes-head Alley out of Cornhill, 1662.

________. *The Doctrine of the Blessed Trinity Briefly Explained to a Friend.* London: Printed for Tho. Parkhurst, at the Bible and Three Crowns, in Cheapside, 1690.

________. "Dr. Wallis's Letter to the Publisher, concerning a new Musical Discovery; written from Oxford, March 14. 1676/77." *Philosophical Transactions* 12 (23 April 1677):839 42.

________. *An Eighth Letter Concerning the Sacred Trinity; Occasioned by some Letters to him on that Subject.* London: Printed for Tho. Parkhurst at the Bible and Three Crowns at the Lower end of Cheapside near Mercers-Chapel, 1692.

________. *An Explication and Vindication of the Athanasian Creed, In a Third Letter, Pursuant of Two former, Concerning the Sacred Trinity. Together With a Postscript, in Answer to another Letter.* London: Printed for Tho. Parkhurst, at the Bible and Three Crowns, in Cheapside, 1691.

________. *A Fifth Letter, Concerning the Sacred Trinity; in Answer To what is Entituled, The Arian's Vindication of himself against Dr. Wallis's Fourth Letter on the Trinity.* London: Printed for Tho. Parkhurst, at the Bible and Three Crowns, in Cheapside, 1691.

________. *A Fourth Letter, Concerning the Sacred Trinity; in Reply To what is Entituled, An Answer To Dr. Wallis's Three Letters.* London: Printed for Tho. Parkhurst, at the Bible and Three Crowns, in Cheapside, 1691.

________. *Institutio Logicae. Ad communes usus acommodata.* Oxford: E Theatro Sheldoniano, Prostant apud Amos Curteyne, 1687.

________. "John Wallis to Thomas Smith, 29 January 1696/97." In *The Works of Thomas Hearne.* Vol. 3: *Peter Langtoft's Chronicle*, pp. cxl-clxx. [London]: Printed for Samuel Bagster, 1810.

________. *The Life of Faith. In Two Sermons to the University of Oxford, At St. Mary's Church there; On the 6th. of January, 1683/4 and June the 29th. following.* London: Printed by James Rawlins, for Thomas Parkhurst: And are to be sold by Amos Curteine, Bookseller in Oxford, 1684.

________. *The Resurrection Asserted: In a Sermon Preached to the University of Oxford, on Easter-Day, 1679.* Oxford: Printed by Hen: Hall for James Good, 1679.

________. *A second letter Concerning the Holy Trinity. Pursuant to the former from the same Hand; Occasioned by a Letter (There inserted) from one unknown.* London: Printed for Tho. Parkhurst at the Bible and Three Crowns in Cheapside near Mercers-Chapel, 1691.

________. *A Seventh Letter, Concerning the Sacred Trinity; Occasioned by a Second Letter from W. J.* London: Printed for Tho. Parkhurst, at the Bible and Three Crowns, in Cheapside, 1691.

________. *A Sixth Letter, Concerning the Sacred Trinity; In Answer To a Book Entituled, Observations on the Four Letters, &c.* London: Printed for Tho. Parkhurst, at the Bible and Three Crowns, in Cheapside, 1691.

________. *Three Sermons Concerning the Sacred Trinity.* London: Printed for Tho. Parkhurst, at the Bible and Three Crowns, in Cheapside, 1691.

________. *A Treatise of Algebra, Both Historical and Practical. Shewing, The Original, Progress, and Advancement thereof, from time to time; and by what Steps it hath attained to the Heighth at which now it is. With some Additional Treatises, I. Of the Cono-Cuneus; being a Body representing in part a Conus, in part a Cuneus. II. Of Angular Sections; and other things relating thereunto, and to Trigonometry. III. Of the Angle of Contact; with other things appertaining to the Composition of Magnitudes, the Inceptives of Magnitudes, and the Compositions of Motions, with the results thereof. IV. Of Combinations, Alternations, and Aliquot Parts.* London: Printed by John Playford, for Richard Davis, Bookseller, in the University of Oxford, 1685.

[Walker, John, ed.]. *Letters Written by Eminent Persons in the Seventeenth and Eighteenth Centuries: To Which Are Added, Hearne's Journeys to Reading, and to Whaddon Hall, the Seat of Browne Willis, Esq., and Lives of Eminent Men, by John Aubrey, Esq. The Whole Now First Published From the Originals in the Bodleian Library and Ashmolean Museum, with Biographical and Literary Illustrations.* 3 vols. London: Longman, Hurst, Rees, Orme, and Brown, Paternoster-row; and Munday and Slatter, Oxford, 1813.

Ward, Seth. *An Apology for The Mysteries of the Gospel. Being A Sermon Preached at White-Hall, Feb. 16 1672/3. Printed by his Majesties special Command.* London: Printed by Andrew Clark, for James Collins, 1673.

________. *A Philosophicall Essay Towards an Eviction of The {Being and Attributes of God. {Immortality of the souls of men. {Truth and Authority of Scripture. Together With an Index of the Heads of every particular Part.* 2nd. ed. Oxford: Printed by Leonard Lichfield, and are to be sold by Edward Forrest, 1655.

________. *A Sermon Preached before the Peers, in the Abby-Church at Westminster: October 10th, M. DC. LXVI.* London: Printed for E. C. for James Collins, and are to be sold at the Kings-bead in Westminster-Hall, 1666.

________. *Concerning the Sinfulness, Danger & Remedies of Infidelity. A Sermon Preached at Whitehall, February 16. 1667/68.* London: Printed by A. C. for James Collins, at the Kings Arms within Ludgate near St. Pauls, 1672. In *Six Sermons Preached By the Right Reverend Father in God, Seth Lord Bishop of Sarum,* pp. 169-238. London: Printed by Andrew Clark, for James Collins, at the Kings Arms in Ludgate-street, 1672.

[________]. *Vindiciae Academiarum Containing, Some briefe Animadversions upon Mr Websters Book, Stiled, The Examination of Academies. Together with an Appendix concerning what M. Hobbs, and M. Dell have published on this Argument.* Oxford: Printed by Leonard Lichfield Printer to the University, for Thomas Robinson, 1654.

Waterhouse, Edward. *A Discourse and Defence of Arms and Armory, Shewing The Nature and Rises of Arms and Honour in England, from the Camp, the Court, the City: Under The two later of which, are contained Universities and Inns of Court.* London: T.R. for Samuel Mearne in Little Britain, 1660.

________. *Fortescutus Illustratus, or a Commentary On that Nervous Treatise De Laudibus Legum Angliae, Written by Sir John Fortescue Knight, First Lord Chief Justice, after Lord Chancellour to King Henry the Sixth. Which Treatise, dedicated to Prince Edward that King's Son and Heir (Whom he attended in his retirement into France, and to Whom he loyally and affectionately imparted Himself in the Virtue and Variety of His Excellent Discourse) Hee purposely wrote to consolidate his Princely minde in the love and approbation of the good Lawes of England, and of the laudable Customs of this his Native Country.* London: Printed by Tho. Roycroft for Thomas Dicas at the Sign of the Hen and Chickens in St. Paul's Church-yard, 1663.

________. *The Gentlemans Monitor; Or a Sober Inspection into the Vertues, Vices, and Ordinary Means, Of the Rise and Decay of Men and Families. With The Authors Apology and Application to the Nobles and Gentry of England. Seasonable for these Times.* London: Printed for T. R. for R. Royston, Bookseller to his most Sacred Majesty, 1665.

________. *An humble Apologie for Learning and Learned Men.* London: Printed by T. M. for M. M. G. Bedell, and T. Collins, at the middle Temple in Fleetstreet, 1653.

Wetenhall, Edward. *Scripture Authentick, and Faith Certain. A Discourse, Which may serve for an Answer to divers late Aspersions on the Integrity of Originals, and Validity of our modern Translations.* London: Printed for H. Faithorne, and J. Kersey, at the Rose in S. Paul's Church-yard, 1686.

Wheler, Sir George. *An Account of the Churches, or Places of Assembly, of the Primitive Christians; From the Churches of {Tyre, Jerusalem, and Constantinople. Described by Eusebius. And Ocular Observations of Several very Ancient Edifices of Churches yet Extant in those Parts. With A Seasonable Application.* London: Printed by S. Roycroft, for R. Clavell at the Peacock, at the West-End of S. Pauls, 1689.

________. *A Journey into Greece By George Wheler Esq; In company of Dr Spon of Lyons. In Six Books. Containing I. A Voyage from Venice to Constantinople. II. An Account of Constantinople and the Adjacent Places. III. A Voyage through Lesser Asia. IV. A Voyage from Zant through several Parts of Greece to Athens. V. An Account of Athens. VI. Several Journeys from Athens, into Attica, Corinth, Boetia, &c. With variety of Sculptures.* London: Printed for William Cademan, Robert Kettlewell, and Awnsham Churchill, at the Popes Head in the New Exchange, the Hand and Scepter in Fleetstreet, and the Black Swan near Amen-Corner, 1682.

[Wilkins, John]. *A Discourse concerning A New Planet. Tending to prove, That 'tis probable our Earth is one of the Planets. The second Booke, now first published.* London: Printed by R. H. for Iohn Maynard, and are to be sold at the George in Fleetstreet, neer S. Dunstans Church, 1640. In [John Wilkins], *A Discourse concerning A New world & Another Planet In 2 Bookes*. London: Printed for Iohn Maynard, & are to be sold at the George, in Fleetstreet near St. Dunstans Church, 1640.

[________]. *The Discovery of a World in the Moone. Or, A Discourse Tending, To Prove, that 'tis probable there may be another habitable World in that Planet*. London: Printed by E. G. for Michael Sparke and Edward Forrest, 1638.

[________]. *The Discovery of a New World. Or, A Discourse tending to prove, that 'tis probable there may be another habitable World in the Moone. With a Discourse concerning the possibility of a Passage thither. Third impression. Corrected and enlarged*. 3rd. ed. London: Printed by Iohn Norton for Iohn Maynard, and are to be sold at the George in Fleetstreet, neere St. Dunstons Church, 1640.

________. *An Essay Towards a Real Character And a Philosophical Language*. London: For Sa: Gellibrand, and for John Martyn Printer to the Royal Society, 1668.

________. *Mathematical Magick. Or The Wonders That may be performed by Mechanicall Geometry. In two Books. Concerning Mechanicall {Powers. {Motions. Being One of The most easie, pleasant, usefull, (and yet most neglected) part of Mathematicks. Not before treated of in this language*. London: Printed by M. F. for Sa: Gellibrand at the brasen Serpent in Pauls Church-yard, 1648.

________. *Mercury, or the Secret and Swift Messenger: Showing, How a Man may with Privacy and Speed communicate his Thoughts to a Friend at any distance*. London: Printed by I. Norton, for Iohn Maynard, and Timothy Wilkins, and are to be sold at the George in Fleetstreet, neere Saint Dunstans Church, 1641.

________. *Of the Principles and Duties of Natural Religion: Two Books. To which is added, A Sermon Preached at his Funeral, by William Lloyd, D.D. Dean of Bangor, and Chaplain in Ordinary to His Majesty*. 3rd. ed. London: Printed for T. Basset, R. Chiswell, and C. Brome, at the George in Fleetstreet, the Rose and Crown in St. Pauls Church-Yard, and the Gun at the West-end of St. Pauls, 1693; facsimile reprint ed., with an introduction by Henry G. Van Leeuwen, New York: Johnson Reprint Corp., 1969.

Willughby, Francis. *De Historia Piscium Libri Quatuor, Jussu & Sumptibus Societatis Regiae Londinensis editi. In quibus non tantum De Piscibus in genere agitur, Sed & species omnes, tum ab aliis traditae, tum novae & nondum editae bene multae, naturae ductum servante Methodo dispositae, accurate describuntur. Earumque effigies, quotquot haberi potuere, vel ad vivum delineatae, vel ad optima exemplaria impressae; Artifici manu elegantissime in AEs incisae, ad descriptiones illustrandas exhibentur. Cum Appendice Historias & Observationes supplementum Operis collatas complectente*. Edited by John Ray. Oxford: E Theatro Sheldoniano, 1686.

________. *Ornithologiae Libri Tres: In quibus Aves omnes hactenus cognitae in methodum naturis suis convenientem redactae accurate describuntur, Descriptiones Iconibus elegantissimis & vivarum Avium simillimis, AEri incisis illustrantur*. Edited by John Ray. London: Impensis Joannis Martyn, Regiae Societatis Typographi, ad insigne Campanae in Coemeterio D. Pauli, 1676.

Wood, Anthony à. *Athenae Oxonienses. An Exact History of All the Writers and Bishops Who Have Had Their Education in the University of Oxford. To Which Are Added, The Fasti, or Annals of the Said University*. Edited by Philip Bliss. 3rd. ed. 4 vols. London: Printed for F. C. and J. Rivington; Lackington, Allen, and Co.; T. Payne;

White, Cochrane, and Co.; Longman, Hurst, Rees, Orme, and Brown; Cadell and Davies; J. and A. Arch; J. Mawman; Black, Parry, and Co.; R. H. Evans; J. Booth, London; and J. Parker, Oxford, 1813-20; facsimile reprint ed., Anglistica & Americana, no. 22, Hildesheim: George Olms, 1969.

________. *The History and Antiquities of the Colleges and Halls in the University of Oxford:... Now First Published in English, From the Original Manuscript in the Bodleian Library; With a Continuation to the Present Time.* Edited by John Gutch. Oxford: Clarendon Press, 1786.

________. *The Life and Times of Anthony Wood.* Edited by Andrew Clark. 5 vols. Oxford: Printed for the Oxford Historical Society, at the Clarendon Press, 1891-1900.

W[ood], R[obert]. "TEN TO ONE or A short and ready way For the extraordinary Facilitation & Dispatch of RECKONINGS. By meanes only of Two or Three NEW DECIMAL COINES much desired in order to The Publick Good." In Charles Webster, *The Great Instauration: Science, Medicine and Reform, 1626-1660,* pp. 535-39. New York: Holmes & Meier Publishers, 1976.

Secondary Sources

Aarsleff, Hans. "Leibniz on Locke on Language." *American Philosophical Quarterly* 1 (1964):165-88.

________. "Thomas Sprat." In *Dictionary of Scientific Biography*, 12:580-87. Edited by Charles Coulston Gillispie. 16 vols. New York: Charles Scribner's Sons, 1970-80.

________. "John Wilkins." In *Dictionary of Scientific Biography*, 14:361-81. Edited by Charles Coulston Gillispie. 16 vols. New York: Charles Scribner's Sons, 1970-80.

Andrade, E. N. da C. *A Brief History of The Royal Society.* London: The Royal Society, 1960.

Armstrong, Brian G. *Calvinism and the Amyraut Heresy.* Madison: The University of Wisconsin Press, 1969.

Austin, William H. "Henry More." In *Dictionary of Scientific Biography*, 9:509-10. Edited by Charles Coulston Gillispie. 16 vols. New York: Charles Scribner's Sons, 1970-80.

________. "Joseph Glanvill." In *Dictionary of Scientific Biography*, 5:414-17. Edited by Charles Coulston Gillispie. 16 vols. New York: Charles Scribner's Sons, 1970-80.

Ball, W. W. Rouse, and Venn, J. A. *Admissions to Trinity College, Cambridge.* 5 vols. London: Macmillan and Co., Ltd., 1913.

Baron, Margaret E. *The Origins of the Infinitesimal Calculus.* Oxford: Pergamon Press, 1969.

Blake, Ralph M.; Ducasse, Curt J.; and Madden, Edward H. *Theories of Scientific Method: The Renaissance through the Nineteenth Century.* Seattle: University of Washington Press, 1960.

Bluhm, R. K. "Remarks on the Royal Society's Finances." *Notes and Records of the Royal Society of London* 13 (November 1958):82-103.

Bosher, Robert S. *The Making of the Restoration Settlement: The Influence of the Laudians, 1649-1662.* London: Dacre Press, 1951.

Brody, B. A. "Towards an Aristotelean Theory of Scientific Explanation." *Philosophy of Science* 39 (March 1972):20-31.

Burtt. E. A. *The Metaphysical Foundations of Modern Science.* Garden City, NY: Doubleday & Co., Anchor Books, 1954.

Byrne, Edmund F. *Probability and Opinion.* The Hague: Martinus Nijhoff, 1968.

Cambridge University. *Cantabrigienses Graduati sive Catalogus exhibens nomina eorum, Quos ab anno 1659, usque ad annum 1787, gradu quocunque ornavit Academia Cantabrigiensis; e libris subscriptionum desumptus, atque ordine alphabetico compositus.* Cambridge: Typis Academicis Excudebat J. Archdeacon. Veneunt apud J. & J. Merrill, Cantabrigiae; et T. Cadell, T. Payne & Fil. B. White & Fil. et G. & G. Wilkie, Londini, 1787.

Carroll, Robert Todd. *The Common-sense Philosophy of Religion of Bishop Edward Stillingfleet, 1635-1699.* The Hague: Martinus Nijhoff, 1975.

Cassirer, Ernst. *The Platonic Renaissance in England.* Translated by James P. Pettegrove. Austin: University of Texas Press, 1953; reprint ed., New York: Gordian Press, 1970.

Charlton, Kenneth. *Education in Renaissance England.* London: Routledge and Kegan Paul, 1965.

Child, J. M. Introduction to *The Geometrical Lectures of Isaac Barrow.* Translated and Edited by J. M. Child. The Open Court Series of Classics of Science and Philosophy, no. 3. Chicago: The Open Court Publishing Company, 1916.

Chisholm, Roderick M. *Theory of Knowledge.* 2nd. ed. Englewood Cliffs, New Jersey: Prentice-Hall, Inc., 1977.

Clark, Andrew. *Register of the University of Oxford.* 2 vols. Oxford: Oxford Historical Society, 1885-89.

Cohen, I. Bernard. *Franklin and Newton.* Philadelphia: The American Philosophical Society, 1956.

Cooper, Charles Henry. *Annals of Cambridge.* 5 vols. London: Warwick & Co., 1842-53.

Cope, Edward Meredith. *An Introduction to Aristotle's Rhetoric.* London: Macmillan and Co., 1867; reprint ed., Dubuque IA: William C. Brown Reprint Library, n. d.

Cope, Jackson I. *Joseph Glanvill, Anglican Apologist.* St. Louis: Washington University Studies, 1956.

Costello, William T. *The Scholastic Curriculum at Early Seventeenth-Century Cambridge.* Cambridge: Harvard University Press, 1958.

Cragg, G. R. *From Puritanism to the Age of Reason.* Cambridge: Cambridge University Press, 1966.

Curtis, Mark H. *Oxford and Cambridge in Transition, 1558-1642.* Oxford: Clarendon Press, 1959.

Danto, Arthur Coleman. *Analytical Philosophy of Action.* Cambridge: Cambridge University Press, 1973.

Davidson, Donald. "On the Very Idea of a Conceptual Scheme." *Proceedings and Addresses of the American Philosophical Association* 47 (1973-74):5-20.

________. "Psychology as Philosophy." In *The Philosophy of Mind*, pp. 101-10. Edited by Jonathan Glover. Oxford Readings in Philosophy. New York: Oxford University Press, 1976.

Davies, Horton. *Worship and Theology in England.* Vol. 2: *From Andrewes to Baxter and Fox, 1603-1690.* Princeton: Princeton University Press, 1975.

Davis, Lawrence H. *Theory of Action.* Englewood Cliffs, N.J.: Prentice-Hall, Inc., 1979.

De Beer, E. S. "The Earliest Fellows of the Royal Society." *Institute of Historical Research Bulletin* 15 (1937):79-93.

Debus, Allen G. *Science and Education in the Seventeenth Century: The Webster-Ward Debate.* New York: American Elsevier Inc., 1970.

Dictionary of National Biography From the Earliest Times to 1900. 22 vols. Oxford: Oxford University Press, 1885-1900.

Dijksterhuis, E. J. "John Pell in Zijn Strijd Over De Rectificatie van den Cirkel." *Euclides* 8 (1932):286-96.

________. *The Mechanization of the World Picture.* Translated by C. Dikshoorn. Oxford: Clarendon Press, 1961.

Dillenberger, John. *Protestant Thought and Natural Science: A Historical Interpretation.* London: Collins, 1961.

Douglas, David C. *English Scholars, 1660-1730.* 2d revised ed. London: Eyre & Spottiswoode, 1951.

Echard, Laurence. *The History of England. From the First Entrance of Julius Caesar and the Romans, To the Conclusion of the Reign of King James the Second And the Establishment of King William and Queen Mary Upon the Throne, in the Year 1688. With a Compleat Index.* 3rd. ed. London: Printed for Jacob Tonson, at Shakespear's-head over-against Katharine-street in the Strand, 1720.

'Espinasse, Margaret. "The Decline and Fall of Restoration Science." In *The Intellectual Revolution of the Seventeenth Century*, pp. 347-68. Edited by Charles Webster. Boston: Routledge & Kegan Paul, Ltd., 1974.

Evans, J. L. "Error and the Will." *Philosophy* 38 (1963):136-48.

Feyerabend, Paul K. "Classical Empiricism." In *The Methodological Heritage of Newton*, pp. 150-70. Edited by R. E. Butts and J. W. Davies. Toronto: University of Toronto Press, 1970.

Finlayson, Michael G. *Historians, Puritanism, and the English Revolution: The Religious Factor in English Politics Before and After the Interregnum*. Buffalo: University of Toronto Press, 1983.

Finocchiaro, Maurice A. *Galileo and the Art of Reasoning: Rhetorical Foundations of Logic and Scientific Method*. Boston Studies in the Philosophy of Science, vol. 61. Dordrecht: D. Reidel Publishing Co., 1980.

Fortenbaugh, W. W. *Aristotle on Emotion: A Contribution to Philosophical Psychology, Rhetoric, Poetics, Politics and Ethics*. London: Gerald Duckworth & Company Ltd., 1975.

Foster, Joseph. *Alumni Oxonienses: The Members of the University of Oxford, 1500-1714: Their Parentage, Birthplace and Year of Birth, with a Record of their Degrees. Being The Matriculation Register of the University, Alphabetically Arranged, Revised, and Annotated*. 4 vols. Oxford: Parker & Co., 1891-92.

France, Peter. *Rhetoric and Truth in France: Descartes to Diderot*. Oxford: Clarendon Press, 1972.

Frank, Robert Gregg, Jr. *Harvey and the Oxford Physiologists: A Study of Scientific Ideas and Social Interaction*. Berkeley: University of California Press, 1980.

________. "Institutional Structure and Scientific Activity in the Early Royal Society." In *Proceedings No. 4 of the XIV International Congress of the History of Science*, pp. 82-99. Tokyo: n. p., 1975.

________. "Science, Medicine and the Universities of Early Modern England: Background and Sources." *History of Science* 11 (1973):194-216, 239-69.

Gardiner, Patrick. "Error, Faith, and Self-deception." In *The Philosophy of Mind*, pp. 35-52. Edited by Jonathan Glover. Oxford Readings in Philosophy. New York: Oxford University Press, 1976.

Gardiner, R. B. *The Registers of Wadham College, Oxford. From 1613-1871*. 2 vols. London: G. Bell and Sons, 1889-95.

Gillispie, Charles Coulston. *Dictionary of Scientific Biography*. 16 vols. New York: Charles Scribner's Sons, 1970-80.

Gilson, Etienne Henry. *Reason and Revelation in the Middle Ages*. New York: Charles Scribner's Sons, 1938.

Grant, Edward. "Hypotheses in Late Medieval and Early Modern Science." *Daedalus* 91 (1962):599-612.

Green, I. M. *The Re-establishment of the Church of England: 1660-1663.* New York: Oxford University Press, 1978.

Greene, R. A. "Henry More and Robert Boyle on the spirit of nature." *Journal of the History of Ideas* 23 (October-December 1962):451-74.

Gunther, Robert W. T. *Early Science in Cambridge.* Oxford: Printed for the author at the University Press, 1937.

________. *Early Science in Oxford.* Vol. 11: *Oxford Colleges and Their Men of Science.* Oxford: Printed for the author, 1937.

Hacking, Ian. *The Emergence of Probability: A Philosophical Study of Early Ideas about Probability, Induction and Statistical Inference.* New York: Cambridge University Press, 1975.

________. "Leibniz and Descartes: Proof and Eternal Truths." *Proceedings of the British Academy* 59 (1973):175-88.

________. *Why Does Language Matter to Philosophy?* Cambridge: Cambridge University Press, 1975.

Hahn, Roger. *The Anatomy of a Scientific Institution: The Royal Academy of Sciences 1666-1803.* Berkeley: University of California Press, 1971.

Hall, Alfred Rupert. "Merton Revisited or Science and Society in the Seventeenth Century." *History of Science* 2 (1963):1-16.

________. *The Scientific Revolution, 1600-1800.* London: Longmans, Green and Co., 1954.

Hall, Marie Boas. "Science in the Early Royal Society." In *The Emergence of Science in Western Europe*, pp. 57-77. Edited by Maurice P. Crosland. New York: The Macmillan Press, Ltd., 1975.

Hartley, Sir Harold. *The Royal Society: Its Origins and Founders.* London: The Royal Society of London, 1960.

Hawks, Ellison. *Pioneers of Plant Study.* n.p., 1928; reprint ed., Freeport, NY: Books for Libraries Press, Essay Index Reprint Series, 1969.

Hick, John. *Faith and Knowledge.* 2nd. ed. Ithaca: Cornell University Press, 1966.

Hill, Abraham. "Some Account of the Life of Dr. Isaac Barrow: To the Rev. Dr. Tillotson, Dean of Canterbury." In *The Theological Works of Isaac Barrow, D.D.*, 1:xxxvii-lvi. Edited by Alexander Napier. 9 vols. Cambridge: Cambridge University Press, 1859.

Hill, Christopher. *Intellectual Origins of the English Revolution.* London: Panther Books, 1972.

________. "The Intellectual Origins of the Royal Society, London or Oxford?" *Notes and Records of the Royal Society of London* 23 (1968):144-56.

________. "Puritanism, Capitalism and the Scientific Revolution." In *The Intellectual Revolution of the Seventeenth Century*, pp. 243-53. Edited by Charles Webster. Boston: Routledge & Kegan Paul, Ltd., 1974.

Hintikka, Jaakko. "On the Ingredients of an Aristotelian Science." *Nous* 6 (1972):55-69.

Hirschberg, Daniel Ray. "A Social History of the Anglican Episcopate." Ph.D. dissertation, University of Michigan, 1976; Ann Arbor, MI: University Microfilms, DCJ76-27493, n. d.

Hoppen, K. Theodore. *The Common Scientist of the Seventeenth Century: A Study of the Dublin Philosophical Society, 1683-1708*. Charlottesville: University Press of Virginia, 1970.

Houghton, Walter E. "The English Virtuoso in the Seventeenth Century." *Journal of the History of Ideas* 3 (1942):51-73, 190-219.

Howell, Wilbur Samuel. *Logic and Rhetoric in England, 1500-1700*. New York: Russell & Russell, Inc., 1961.

Hunter, Michael Cyril William. *John Aubrey and the Realm of Learning*. London: Gerald Duckworth & Co. Ltd., 1975.

________. *Science and Society in Restoration England*. Cambridge: Cambridge University Press, 1981.

________. "The Social Basis and Changing Fortunes of an Early Scientific Institution: An Analysis of the Membership of the Royal Society, 1660-1685." *Notes and Records of the Royal Society of London* 31 (July 1976):9-114.

Jacob, James R. *Henry Stubbe: radical Protestantism and the early Enlightenment*. Cambridge: Cambridge University Press, 1983.

________. "Restoration Ideologies and the Royal Society." *History of Science* 18 (March 1980):25-38.

________. "Restoration, Reformation, and the Origins of the Royal Society." *History of Science* 13 (1975):155-76.

________. *Robert Boyle and the English Revolution*. New York: Burt Franklin & Co. Inc., 1977.

Jacob, James R., and Jacob, Margaret Candee. "Science and Religion." *History of Science* 14 (1976):196-207.

________. "Scientists and Society: The Saints Preserved." *Journal of European Studies* 1 (1971):87-92.

Jacob, Margaret Candee. "The Church and the Foundation of the Newtonian World View." *Journal of European Studies* 1 (1971):128-48.

________. *The Newtonians and the English Revolution 1689-1720*. Ithaca: Cornell University Press, 1976.

Jardine, Lisa. *Francis Bacon, Discovery and the Art of Discourse*. Cambridge: Cambridge University Press, 1974.

Jones, Richard Foster. *Ancient and Moderns: A Study of the Rise of the Scientific Movement in Seventeenth-Century England.* Washington University Studies. 2nd. ed. St. Louis: 1961.

________. "The Rhetoric of Science in England of the Mid-Seventeenth Century." In *Restoration and Eighteenth-Century Literature: Essays in Honor of Alan Dugald McKillop*, pp. 5-24. Edited by Carroll Camden. Chicago: The University of Chicago Press for the William Marsh Rice University, 1963.

________. *The Seventeenth Century: Studies in the History of English Thought and Literature from Bacon to Pope*. Palo Alto: Stanford University Press, 1951.

Kargon, Robert. *Atomism in England From Hariot to Newton.* Oxford: Clarendon Press, 1966.

________. "Newton, Barrow and the Hypothetical Physics." *Centaurus* 11 (1965):46-56.

Kearney, Hugh F. "Merton Revisited." *Science Studies* 3 (January 1973):72-78.

________. "Puritanism and Science: Problems of Definition." In *The Intellectual Revolution of the Seventeenth Century*, pp. 254-61. Edited by Charles Webster. Boston: Routledge & Kegan Paul, Ltd., 1974.

________. *Scholars and Gentlemen: Universities and Society in Pre-Industrial Britain, 1500-1700*. London: Faber and Faber, Ltd., 1970.

Kelly, J. N. D. *Early Christian Creeds*. 3rd. ed. New York: David McKay Company Inc., 1972.

Kennedy, G. A. *The Art of Persuasion in Greece.* Princeton: Princeton University Press, 1963.

Keynes, Sir Geoffrey. *John Ray: A Bibliography 1660-1970*. Amsterdam: Gerard Th. Van Heusden, 1976.

Kemsley, Douglas S. "Religious Influence in the Rise of Modern Science: A Review and Criticism, Particularly of the 'Protestant-Puritan Ethic' Theory." *Annals of Science* 24 (1968):199-226.

Kenny, Anthony John Patrick. *Faith and Reason.* Bampton Lectures in America, no. 22. New York: Columbia University Press, 1983.

________. "Freedom, spontaneity and indifference." In *Essays on Freedom of Action*, pp. 89-104. Edited by Ted Honderich. Boston: Routledge & Kegan Paul, Ltd., 1973.

[Kippis, Andrew, ed.]. *BIOGRAPHIA Britannica: or, the Lives of the Most Eminent Persons Who have flourished in Great Britain and Ireland, From the earliest Ages down to the present Times; Collected from the best Authorities both Printed and Manuscript And digested in the Manner of Mr. Bayle's Historical and Critical Dictionary*. 7 vols. London: Printed for W. Meadows, J. Walthoe, T. Osborne, Dr. Browne, H. Whitridge, C. Hitch and L. Hawes, R. Manby, J. & R. Tonson, John Rivington, C. Bathurst, J. Rivington and J. Fletcher, L. Davis and C. Reymers,

J. Ward, R. Baldwin, W. Johnston, J. Richardson, T. Longman, P. Davey and B. Law, T. Kearsley, and M. Cooper, 1747-66.

Koyre, Alexander. *From the Closed World to the Infinite Universe.* Baltimore: The Johns Hopkins Press, 1957.

Kubrin, David. "Newton and the Cyclical Cosmos: Providence and the Mechanical Philosophy." *Journal of the History of Ideas* 28 (1967):325-46.

Kuhn, Thomas S. *The Structure of Scientific Revolutions.* International Encyclopedia of Unified Science, Vol. 2, No. 2. 2d. ed. Chicago: University of Chicago Press, 1970.

Laudan, Laurens. "The Clock Metaphor and Probabilism: The Impact of Descartes on English Methodological Thought, 1650-65." *Annals of Science* 22 (June 1966):73-104.

________. "Method and the Mechanical Philosophy." *History of Science* 5 (1966):117-24.

________. "The Nature and Sources of Locke's Views on Hypotheses." *Journal of the History of Ideas* 28 (1967):211-23.

Le Neve, John. *Fasti Ecclesiae Anglicanae.* 4 vols. The Institute of Historical Research. London: The Athlone Press, 1969-75.

Lichtenstein, Aharon. *Henry More: The Rational Theology of a Cambridge Platonist.* Cambridge: Harvard University Press, 1962.

Lloyd, G. E. R. *Aristotle: The Growth and Structure of His Thought.* Cambridge: Cambridge University Press, 1968.

Lyons, Sir Henry. *The Royal Society 1660-1940.* Cambridge: Cambridge University Press, 1944.

McAdoo, H. R. *The Spirit of Anglicanism: A Survey of Anglican Theological Method in the Seventeenth Century.* London: Adam & Charles Black, Ltd., 1965.

McGuire, J. E. "Force, Active Principles, and Newton's Invisible Realm." *Ambix* 15 (1968):154-208.

________. "Neoplatonism and Active Principles: Newton and the Corpus Hermeticum." In *Hermeticism and the Scientific Revolution*, pp. 95-142. Edited by Robert S. Westman and J. E. McGuire. Los Angeles: William Andrews Clark Memorial Library, University of California, 1977.

McGuire, J. E., and Rattansi, P. M. "Newton and the 'Pipes of Pan.'" *Notes and Records of the Royal Society of London* 21 (1966):108-43.

MacIntyre, Alisdair. "Epistemological Crises, Dramatic Narrative and the Philosophy of Science." *Monist* 60 (1977):453-72.

Mckie, Douglas. "The Origins and Foundation of the Royal Society of London." *Notes and Records of the Royal Society* 15 (1960):1-37.

Madden, Edward H. "The Enthymeme: Crossroads of Logic, Rhetoric, and Metaphysics." *Philosophical Review* 61 (July 1952):368-77.

Mahoney, Michael Sean. *The Mathematical Career of Pierre de Fermat (1601-1665)*. Princeton: Princeton University Press, 1973.

Mandelbaum, Maurice H. *Philosophy, Science, and Sense Perception: Historical and Critical Studies*. Baltimore: The Johns Hopkins University Press, 1964.

Manuel, Frank. *Isaac Newton, Historian*. Cambridge: Harvard University Press, The Belknap Press, 1963.

Matthews, A. G. *Calamy Revised*. Oxford: Clarendon Press, 1934.

________. *Walker Revised*. Oxford: Clarendon Press, 1948.

Merton, Robert K. "Puritanism, Pietism and Science." In *Science and Religious Belief: A Selection of Recent Historical Studies*. Edited by Colin Archibald Russell. London: Open University Press, University of London Press, Ltd., 1973.

________. *Science, Technology, and Society in Seventeenth Century England*. New York: Howard Fertig, 1970.

Moore, J. T. "Locke On Assent and Toleration." *The Journal of Religion* 58 (1978):30-36.

________. "Locke's Analysis of Language and the Assent to Scripture." *Journal of the History of Ideas* 37 (1976):707-14.

Mulligan, Lotte. "Anglicanism, Latitudinarianism and Science in Seventeenth-Century England." *Annals of Science* 30 (June 1973):213-19.

________. "Civil War Politics, Religion and the Royal Society." In *The Intellectual Revolution of the Seventeenth Century*, pp. 317-339. Edited by Charles Webster. Boston: Routledge & Kegan Paul, Ltd., 1974.

Mullinger, James Bass. *The University of Cambridge from the Earliest Times*. 3 vols. Cambridge: Cambridge University Press, 1873-1911.

Nauert, Charles G., Jr. "The Clash of Humanists and Scholastics: An Approach to Pre-Reformation Controversies." *The Sixteenth Century Journal* 4 (April 1973):1-18.

Nelson, Benjamin. "'Probabilists,' 'Anti-Probabilists,' and the Quest for Certitude in the 16th and 17th Centuries." *Proceedings of the Tenth International Congress of the History of Science, Ithaca 1962*, pp. 269-73. Paris: Hermann, 1964.

O'Connor, Daniel and Oakley, Francis. *Creation: The Impact of an Idea*. New York: Charles Scribner's Sons, 1969.

O'Day, Rosemary. *The English Clergy: The Emergence and Consolidation of a Profession, 1558-1642*. Leicester: Leicester University Press, 1979.

O'Day, Rosemary, and Heal, Felicity. *Princes & Paupers in the English Church, 1500-1800*. Leicester: Leicester University Press, 1981.

Orr, Robert R. *Reason and Authority: The Thought of William Chillingworth*. Oxford: Clarendon Press, 1967.

Osler, Margaret J. "Locke and the Changing Ideal of Scientific Knowledge." *Journal of the History of Ideas* 31 (1970):3-16.

Osmond, Percy H. *Isaac Barrow: His Life and Times*. London: S.P.C.K., 1944.

Ozment, Stephen E. *The Age of Reform 1250-1550*. New Haven: Yale University Press, 1980.

Pattison, Mark. "Tendencies of Religious Thought in England, 1688-1750." In *Essays and Reviews*, pp. 254-329. London: John W. Parker and Son, West Strand, 1860; facsimile reprint ed., Westmead, Farnborough, Hants: Gregg International Publishers Ltd., 1970.

Peile, John, comp. *Biographical Register of Christ's College, 1505-1905*. 2 vols. Cambridge: Cambridge University Press, 1910-1913.

Popkin, Richard H. *The History of Scepticism from Erasmus to Spinoza*. Berkeley: University of California Press, 1979.

________. "Skepticism, Theology and the Scientific Revolution in the Seventeenth Century." In *Proceedings of the International Colloquium in the Philosophy of Science, London, 1965*. Vol. 3: *Problems in the Philosophy of Science*, pp. 1-28. Edited by Imre Lakatos and Alan Musgrave. Amsterdam: North-Holland Publishing Co., 1968.

Purver, Margery. *The Royal Society: Concept and Creation*. London: Routledge and Kegan Paul, 1967.

Purver, Margery, and Bowen, E. J. *The Beginning of the Royal Society*. Oxford: Clarendon Press, 1960.

Quine, W. V., and Ullian, J. S. *The Web of Belief*. New York: Random House, Inc., 1970.

Rattansi, P. M. "Science and Religion in the Seventeenth Century." In *The Emergence of Science in Western Europe*, pp. 79-85. Edited by Maurice P. Crosland. New York: The Macmillan Press, Ltd., 1975.

________. "The Social Interpretation of Science in the Seventeenth Century." *In Science and Society, 1600-1900*, pp. 1-32. Edited by Peter Mathias. Cambridge: Cambridge University Press, 1972.

Raven, Charles E. *John Ray, Naturalist: His Life and Works*. Cambridge: Cambridge University Press, 1942.

Redwood, John. *Reason, Ridicule and Religion: The Age of Enlightenment in England 1660-1750*. Cambridge: Harvard University Press, 1976.

Reedy, Gerard. *The Bible and Reason: Anglicans and Scripture in Late Seventeenth-Century England*. Philadelphia: University of Pennsylvania Press, 1985.

Rogers, G. A. J. "Boyle, Locke, and Reason." *Journal of the History of Ideas* 27 (April-June 1966):205-16.

Ross, W. D. *Aristotle*. New York: Meridian Books, Inc., 1959.

Rossi, Paolo. *Francis Bacon, From Magic to Science.* Translated by Sacha Rabinovitch. Chicago: The University of Chicago Press, 1968.

________. "Hermeticism, Rationality and the Scientific Revolution." In *Reason, Experiment, and Mysticism in the Scientific Revolution,* pp. 247-73. Edited by M. L. Righinni Bonelli and William R. Shea. New York: Science History Publications, 1975.

Royal Society of London. *The Record of The Royal Society of London.* 3rd. ed. London: Oxford University Press, 1912.

________. *The Signatures in the First Journal-Book and the Charter-Book of the Royal Society: Being a Facsimile of the Signatures of the Founders, Patrons and Fellows of the Society from the Year 1660 down to the Present Time.* London: Printed for the Royal Society at the Oxford University Press, 1936.

Russell, J. L. "Kepler's Laws of Planetary Motion: 1609-1666." *The British Journal for the History of Science* 2 (1964):1-24.

Scala, Gail Ewald. "An Index of Proper Names in Thomas Birch, *The History of the Royal Society* (London, 1756-1757)." *Notes and Records of the Royal Society of London* 28 (April 1974):263-329.

Schmitt, Charles B. *Cicero Scepticus: A Study of the Influence of the Academia in the Renaissance.* The Hague: Martinus Nijhoff, 1972.

Scriba, Christoph J. "The Autobiography of John Wallis." *Notes and Records of the Royal Society of London* 25 (1970):17-46.

________. "John Wallis." In *Dictionary of Scientific Biography,* 14:146-155. Edited by Charles Coulston Gillispie. 16 vols. New York: Charles Scribner's Sons, 1970-80.

Seaver, Paul S. *The Puritan Lectureships: The Politics of Religious Dissent, 1560-1662.* Palo Alto: Stanford University Press, 1970.

Seigel, Jerrold E. *Rhetoric and Philosophy in Renaissance Humanism.* Princeton: Princeton University Press, 1968.

Shaffer, Jerome A. *Philosophy of Mind.* Englewood Cliffs, N.J.: Prentice-Hall, Inc., 1968.

Shapiro, Barbara J. *John Wilkins, 1614-1672.* Berkeley: University of California Press, 1969.

________. "Latitudinarianism and Science in Seventeenth-Century England." In *The Intellectual Revolution of the Seventeenth Century,* pp. 286-316. Edited by Charles Webster. Boston: Routledge & Kegan Paul, Ltd., 1974.

________. "Law and Science in Seventeenth-Century England." *Stanford Law Review* 21 (1969):727-66.

________. *Probability and Certainty in Seventeenth-Century England: A Study of the Relationships Between Natural Science, Religion, History, Law, and Literature.* Princeton: Princeton University Press, 1983.

________. "The Universities and Science in Seventeenth Century England." *Journal of British Studies* 10 (1971):47-82.

Sharp, Andrew. "Edward Waterhouse's View of Social Change in Seventeenth-Century England." *Past & Present* 62 (1974):27-46.

Simon, Irene. *Three Restoration Divines: Barrow, South, and Tillotson.* Bibliotheque de la Faculte de Philosophie et Lettres de l'Universite de Liege, Fascicule 181. 2 vols. Paris: Societe d'Edition "Les Belles Lettres," 1967.

Simon, Joan. *Education and Society in Tudor England.* Cambridge: Cambridge University Press, 1966.

Skinner, Quentin. "Thomas Hobbes and the Nature of the Early Royal Society." *The Historical Journal* 12 (1969):217-39.

Solmsen, Friedrich. "The Aristotelian Tradition in Ancient Rhetoric." *American Journal of Philology* 62 (January-April 1941):35-50, 169-90.

Sorabji, Richard. *Necessity, Cause, and Blame: Perspectives on Aristotle's Theory.* Ithaca: Cornell University Press, 1980.

Spiller, Michael R. G. *"Concerning Natural Experimental Philosophie": Meric Casaubon and the Royal Society.* Boston: Martinus Nijhoff Publishers, 1980.

Spitz, Lewis William. *The Religious Renaissance of the German Humanists.* Cambridge: Harvard University Press, 1963.

Stimson, H. Dorothy. "Puritanism and the New Philosophy in Seventeenth-Century England." *Bulletin of the Institute of the History of Medicine* 3 (May 1935):321-34.

________. *Scientist and Amateurs: A History of the Royal Society.* London: Sigma Books Ltd., 1949; reprint ed., New York: Greenwood Press, 1968.

Stone, Lawrence. *The University in Society.* 2 vols. Princeton: Princeton University Press, 1974.

Stough, Charlotte L. *Greek Skepticism: A Study in Epistemology.* Berkeley: University of California Press, 1969.

Stout, Jeffrey. *The Flight from Authority: Religion, Morality, and the Quest for Autonomy.* Notre Dame: University of Notre Dame Press, 1981.

Strong, E. W. "Barrow and Newton." *Journal of the History of Philosophy* 8 (1970):156-72.

Struik, Dirk Jan, ed. *A Source Book in Mathematics, 1200-1800.* Cambridge: Harvard University Press, 1969.

Sullivan, Robert E. *John Toland and the Deist Controversy: A Study in Adaptations.* Cambridge: Cambridge University Press, 1982.

Syfret, R. H. "Some Early Critics of the Royal Society." *Notes and Records of the Royal Society of London* 8 (October 1950):20-64.

________. "Some Early Reactions to the Royal Society." *Notes and Records of the Royal Society of London* 7 (April 1950):207-58.

Sykes, Norman. *From Sheldon to Secker: Aspects of English Church History, 1660-1768*. The Ford Lectures [1958]. Cambridge: Cambridge University Press, 1959.

Thoren, Victor E. "John Flamsteed." In *Dictionary of Scientific Biography*, 5:22-26. Edited by Charles Coulston Gillispie. 16 vols. New York: Charles Scribner's Sons, 1970-80.

________. "Seth Ward." In *Dictionary of Scientific Biography*, 14:177-78. Edited by Charles Coulston Gillispie. 16 vols. New York: Charles Scribner's Sons, 1970-80.

Tulloch, John. *Rational Theology and Christian Philosophy in England in the Seventeenth Century*. 2d ed. 2 vols. Edinburgh: William Blackwood and Sons, 1874.

Tyrell, Francis Martin. *The Role of Assent in Judgment: A Thomistic Study*. Ph.D. dissertation, Catholic University of America, no. 100. Washington: Catholic University of America Press, 1948.

Van Leeuwen, Henry G. *The Problem of Certainty in English Thought: 1630-90*. The Hague: Martinus Nijhoff, 1963.

Venn, John. *Biographical History of Gonville and Caius College, 1349-1897: Containing a List of All Known Members of the College from the Foundation to the Present Time, with Biographical Notes*. 4 vols. Cambridge: Cambridge University Press, 1897-1912.

Venn, John, and Venn, J. A. *Alumni Cantabrigiensis: A Biographical List of All Known Students, Graduates and Holders of Office at the University of Cambridge from the Earliest Times to 1900*. 4 vols. Cambridge: Cambridge University Press, 1922-54.

________. *The Book of Matriculations and Degrees: a Catalogue of those who have been Matriculated or been admitted to any Degree in the University of Cambridge from 1544 to 1659*. Cambridge: Cambridge University Press, 1913.

Wallis, P. J. "John Pell." In *Dictionary of Scientific Biography*, 10:495-96. Edited by Charles Coulston Gillispie. 16 vols. New York: Charles Scribner's Sons, 1970-80.

Ward, John. *The Lives of the Professors of Gresham College: To which is prefixed The Life of the Founder, Sir Thomas Gresham. With An Appendix, consisting of Orations, Lectures, and Letters, written by the Professors, with other Papers serving to illustrate the Lives*. London: Printed by John Moore in Bartholomew lane for the Author, and sold by W. Innys, J. and P. Knapton, in Ludgate street, F. Gyles in Holbourn, A. Ward in Little Britain, E. Symon in Cornhill, T. Longman in Pater Noster row, J. Noon in Cheapside, R. Hett in the Poultry, A. Millar in the Strand, and J. Stagg in Westminster hall, 1740; facsimile reprint ed., The Sources of Science, no. 71, New York: Johnson Reprint Corporation, 1967.

Webster, Charles. *The Great Instauration: Science, Medicine and Reform, 1626-1650*. New York: Holmes & Meier Publishers, 1976.

________. "Henry More and Descartes: Some New Sources." *The British Journal for the History of Science* 4 (1969):359-77.

________. "Henry Power's Experimental Philosophy." *Ambix* 14 (1967):150-78.

________. "John Ray." In *Dictionary of Scientific Biography*, 11:313-18. Edited by Charles Coulston Gillispie. 16 vols. New York: Charles Scribner's Sons, 1970-80.

________. "The Origins of the Royal Society." *History of Science* 6 (1967):106-28.

________, ed. *The Intellectual Revolution of the Seventeenth Century*. Boston: Routledge & Kegan Paul, Ltd., 1974.

________, ed. *Samuel Hartlib and the Advancement of Learning*. Cambridge: Cambridge University Press, 1970.

Weinstein, Donald. "Interpretations of Renaissance Humanism." *Journal of the History of Ideas* 33 (1972):165-76.

Weld, Charles Richard. *History of the Royal Society*. 2 vols. London: J. W. Parker, 1848; facsimile reprint ed., New York: Arno Press, 1975.

Westfall, Richard S. *Never at Rest: A Biography of Isaac Newton*. Cambridge: Cambridge University Press, 1980.

________. *Science and Religion in Seventeenth-Century England*. New Haven: Yale University Press, 1958.

Whewell, William. "Barrow and His Academical Times As Illustrated in His Latin Works." *The Theological Works of Isaac Barrow, D.D.*, 9:i-lv. Edited by Alexander Napier. 9 vols. Cambridge: Cambridge University Press, 1859.

Whiteside, D. T. "Isaac Barrow." In *Dictionary of Scientific Biography*, 1:473-76. Edited by Charles Coulston Gillispie. 16 vols. New York: Charles Scribner's Sons, 1970-80.

Wiggins, David. "Freedom, Knowledge, Belief and Causality." In *Knowledge and Necessity*, pp. 132-54. Royal Institute of Philosophy Lectures, vol. 3, 1968-69. New York: St. Martin Press, Inc., 1970.

Willey, Basil. *The Seventeenth Century Background: Studies in the Thought of the Age in Relation to Poetry and Religion*. Garden City, NY: Doubleday & Company, Inc., Anchor Books, 1953.

Williams, Bernard Arthur Owen. "Deciding to Believe." In *Problems of the Self: Philosophical Papers 1956-1972*, pp. 136-151. Cambridge: Cambridge University Press, 1973.

Wilson, Curtis A. "From Kepler's Laws, So-called, to Universal Gravitation: Empirical Factors." *Archive for History of Exact Sciences* 6 (1970):89-170.

Winstanley, Denys Arthur. *The University of Cambridge in the Eighteenth Century*. Cambridge: Cambridge University Press, 1958.

________. *Unreformed Cambridge*. Cambridge: Cambridge University Press, 1935; reprint ed., New York: Arno Press, 1977.

Wolfson, Harry Austryn. "The Double Faith Theory in Clement, Saadia, Averroes and St. Thomas, and its Origin in Aristotle and the Stoics." *The Jewish Quarterly Review* 33 (1942-43):213-64.

________. *The Philosophy of the Church Fathers*. 2 vols. Cambridge: Harvard University Press, 1956.

Wood, P. B. "Methodology and Apologetics: Thomas Sprat's *History of the Royal Society*." *The British Journal for the History of Science* 13 (1980):1-26.

Woolhouse, R. S. *Locke's Philosophy of Science and Knowledge*. New York: Barnes & Noble, 1971.

Wordsworth, Christopher. *Scholae Academicae: Some Account of Studies at the English Universities in the Eighteenth Century*. Cambridge: Cambridge University Press, 1877.

INDEX

C

U

V

W

X, Y, Z